Practical Astronomy

Springer

London
Berlin
Heidelberg
New York
Barcelona
Budapest
Hong Kong
Milan
Paris
Santa Clara
Singapore
Tokyo

Other titles in this series

The Observational Amateur Astronomer
Patrick Moore (Ed.)

The Modern Amateur Astronomer
Patrick Moore (Ed.)

Telescopes and Techniques
Chris Kitchin

Small Astronomical Observatories
Patrick Moore (Ed.)

The Art and Science of CCD Astronomy
David Ratledge (Ed.)

The Observer's Year
Patrick Moore

Seeing Stars
Chris Kitchin and Robert W. Forrest

Photo-guide to the Constellations
Chris Kitchin

The Sun in Eclipse
Michael Maunder and Patrick Moore

Software and Data for Practical Astronomers
David Ratledge

Amateur Telescope Making
Stephen F. Tonkin (Ed.)

Observing Meteors, Comets, Supernovae
and other Transient Phenomena
Neil Bone

Astronomical Equipment for Amateurs

Martin Mobberley

Springer

Martin Mobberley, BSc
"Denmara", Cross Green, Cockfield,
Bury St. Edmunds, Suffolk IP30 0LQ, UK

ISBN 1-85233-019-8 Springer-Verlag London Berlin Heidelberg

British Library Cataloguing in Publication Data
Mobberley, Martin Paul
   Astronomical equipment for amateurs. – (Practical
   astronomy)
   1.Astronomical instruments   2.Astronomy – Amateurs' manuals
   I.Title
   522.2
ISBN 1852330198

Library of Congress Cataloging-in-Publication Data
Mobberley, Martin, 1958–
   Astronomical equipment for amateurs / Martin Mobberley.
      p.   cm. – (Practical astronomy)
   ISBN 1–85233–019–8 (alk. paper)
   1. Astronomical instruments–Handbooks, manuals, etc.
2. Astronomical instruments–Amateurs' manuals.   I. Title.
II. Series.
QB86.M62   1998   98–7025
522'.2–dc21   CIP

Typeset by EXPO Holdings, Malaysia
Printed at the Cromwell Press, Trowbridge, Wiltshire
58/3830-543210   Printed on acid-free paper

Preface

When Springer invited me to write a book about Astronomical Equipment, my biggest dilemma was what to put in and what to leave out, especially with the rapid recent growth in CCD technology and the associated PC-based technologies.

There are over 200 astronomical suppliers out there *that I know about* and an equal number of astronomical and photographic gadgets and accessories, and I knew it would be impossible to cover everything in detail. My eventual solution to this dilemma was simple. I asked myself: if I were starting again, from scratch, and was about to buy my first decent telescope, what advice would I have wanted, and which pitfalls would I wish to avoid?

The glossy advertisements in astronomy magazines do not tell the amateur what it is really like to own and use astronomical equipment. I hope this book does.

If I can save a few amateurs from making expensive mistakes by NOT buying the wrong equipment then the book will have been worthwhile.

Astronomy can be a relaxing, visual pastime with a pair of binoculars and a deck-chair. Or it can be a battle against the elements, to clinch that perfect photo of a comet at its peak. I do far more of the latter type of observing, so this book is biased in that direction!

I very much hope you find something in the book to interest you.

March 1998   Martin Mobberley

Acknowledgements

I am indebted to a number of people for their advice, help and the supply of photographs for this book. Firstly, I must thank my parents for their support; in particular, my father's help with the figures has been considerable. I am extremely grateful to Hazel McGee for meticulously reading through the first draft of the book.

I am also indebted to the following amateur astronomers who have kindly donated figures for the book: Mike Collins; Neil Bone; Steve Evans; Henry Soper; Nick James; Mark Armstrong; Henry Hatfield; Maurice Gavin; Gerald North; Andrew Elliott; Glyn Marsh; Denis Buczynski; Terry Platt and Eric Strach.

My thanks must also go to John Watson of Springer, who originally suggested I could write such a book, much to my surprise!

March 1998   Martin Mobberley

Contents

Contents

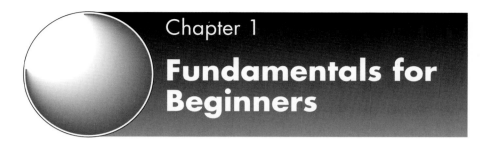

Chapter 1
Fundamentals for Beginners

When – as a schoolboy – I first became interested in amateur astronomy, choosing my first decent telescope was a simple matter; if I saved my pocket money for a year, I could just afford a 60 mm refractor! Saving my pocket money for more than a year was beyond my comprehension and a 60 mm refractor was, to me, a dream telescope, sitting in the window of the local photographic store. To non-astronomers, a telescope *is* a refractor!

Refractors are also the first choice for most young amateur astronomers because small models are rugged, easily available in the high street and lend themselves well to terrestrial use as well as astronomical. So, in reviewing the different types of astronomical telescope, it is only fitting to start with the refractor.

Before we do this however, let us consider a few basic principles.

Using Low Magnification

The purpose of any optical telescope is to collect and focus more light than the human eye can, so that a magnifying eyepiece can be used to inspect the image formed at the focal plane. A large telescope will enable faint objects (stars, nebulae, galaxies and comets) to be seen; it will also enable high magnifications to be used when viewing the Moon and planets (provided the Earth's atmosphere will allow it).

One thing a telescope *cannot* do is to increase the surface brightness of extended faint objects. For example, putting a very low-power (low-magnification) eyepiece in a large-aperture telescope will *not* enable you to see the Orion nebula in the glorious colours seen in astrophotographs by making it look brighter.

Similarly, you will never see the intricate detail in the tails of comets, which can be captured by long-exposure photographs. This limitation may, at first, seem counter-intuitive. After all, if you look at the Andromeda Galaxy (M31) through even a modest pair of binoculars it looks far more impressive than it does with the naked eye.

But this is an illusion. Next time you are looking at M31 on a crystal-clear night and under a dark sky, compare what you see with the naked eye with what you see through the eyepiece. With the naked eye you will see a relatively bright fuzzy stellar centre surrounded by a fainter fuzzy ellipse, maybe several degrees across from a really dark site. Through the telescope that bright centre fills the field and dominates the view, but the extreme outer edges are no easier to trace than they were with the naked eye.

Often, with today's levels of light-pollution from street-lights and similar sources, increasing the brightness with low magnifications may not be very desirable anyway; increasing the contrast using high powers, filters, or by travelling to a dark site is a far better plan.

So – to reiterate – a telescope cannot increase the visual surface brightness of extended objects. Nebulae, galaxies and comets will never look like they do in professional photographs, however large the instrument. Why?

The reason is that the dark-adapted eye has a maximum aperture of about 7 mm (for observers under about thirty years of age). A few young people have exit pupils up to 8 or 9 mm in diameter, but 7 mm is typical. The dark-adapted pupil becomes smaller as you get older, and by the time you are eighty it may have reduced to a mere 3–4 mm maximum. This can become fundamentally important when choosing binoculars, a subject I will talk about later.

The "bundle of light" emerging from a telescope eyepiece (the *exit pupil*, described in the next section) has a diameter equal to the *telescope aperture divided by the magnification.*

A telescope of aperture 70 mm and a magnification of less than 10× will produce an exit pupil too large for even the young human eye to accommodate, so light will be wasted – see Figure 1.1. To avoid wasting light a telescope x times bigger than the aperture of the human eye must use an eyepiece that will give at least x times magnification. (Wasting light may not always be important: you might simply want to use a very low magnification to get a particularly wide field-of-view.) Thus the diameter of the dark-adapted pupil of the human eye sets a fundamental limit on the lowest useful magnification of an optical telescope.

You won't be able to see extended objects in vibrant colours, but a telescope *will* enable you to see far more *detail* in faint nebulous objects. The larger number of photons captured by a large telescope, and its higher resolution, will always pay dividends.

A sensible rule-of-thumb for the *minimum* useful magnification for a (young person's) telescope is one and a half times the aperture in centimetres, e.g. 15× for a 10 cm telescope.

An often unsuspected additional problem occurs with some big reflecting telescopes (see below). If there is a very large secondary mirror obstructing the light path, the resulting shadow spot in the exit pupil also becomes large at low powers and may approach the size of the eye's pupil in extreme cases (e.g. during daylight observing)! This can be a good reason for using refractors for low-power observing.

The shadow spot diameter can be calculated. The formula is:

Shadow spot = (eyepiece exit pupil diameter × secondary mirror diameter) / (primary mirror diameter)

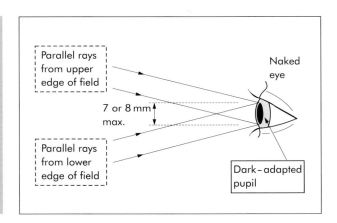

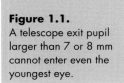

Figure 1.1.
A telescope exit pupil larger than 7 or 8 mm cannot enter even the youngest eye.

Using High Magnification

At the other extreme, high magnification, the limits are largely imposed by the Earth's atmosphere, which rarely allows magnifications greater than 200× to be usefully employed on the Moon and planets. Higher powers can occasionally be useful in increasing the contrast when searching out faint stars or Deep Sky objects.

I don't intend to clutter this book with unnecessary calculations and formulae. However there are a few invaluable formulae which every amateur should know. I have listed the most useful ones below, along with a few explanations of the "jargon".

Formulae

- *f/ratio* = (telescope focal length) / (telescope aperture)
- *magnification* = (telescope focal length) / (eyepiece focal length)
- also, *magnification* = (telescope aperture) / (exit pupil diameter)
- *highest resolution [Dawes limit] of a telescope* = (11.6 arc-seconds) / (telescope aperture in centimetres)
- *image size at the focal plane of a telescope* = $\tan \theta \times$ (telescope focal length)
  [θ is the angular size of the object; the Moon, for example, is 0.5 degrees in diameter]
- *real field of view* = (apparent field of view) / (magnification)

Jargon

Some of the terms in common use may need explanation.

Firstly, the *focal length* of a telescope is simply the distance from the lens or mirror to the focused image; obviously, this will be very similar to the tube length in simple refractors or Newtonian reflectors.

Power is often used as a synonym for magnification.

The *f/ratio* is the focal ratio; the same terminology is used for camera lenses. However, because telescopes have no iris diaphragm to control the amount of light entering the lens, the aperture is rarely "stopped down" (reduced), so the ratio of the mirror or lens diameter to the focal length is usually fixed.

One common misconception (a photographic legacy) is that telescopes of shorter f/ratio give brighter visual images than those of a long f/ratio. As I explained above, the size and surface brightness of an extended object, *when viewed visually*, are dependent on the magnification used and *not* the f/ratio.

The main (largest) lens or mirror is sometimes referred to as the *objective.*

The sizes of astronomical objects are frequently measured in *arc-seconds* or *arc-minutes.* There are 60 arc-minutes in one degree and 60 arc-seconds in one arc-minute; thus the Moon, with a diameter of half a degree, spans 30 arc-minutes or 1800 arc-seconds.

Jupiter, at opposition, is 44 to 50 arc-seconds in diameter. Thus, from the "Dawes limit" formula above, a telescope of 11.6 cm aperture could resolve details separated by 1/44th to 1/50th of the diameter of Jupiter, under perfect atmospheric conditions (see Figure 1.2).

Sometimes the term *diffraction-limited performance* is used in telescope specifications. Essentially, this means the Dawes limit.

Diffraction is the phenomenon that limits the resolution of all instruments that image electromagnetic radiation. The *diffraction limit* is a consequence of the wavelength of the radiation (light, in an optical telescope) and the aperture of the instrument. Shorter

Figure 1.2. Jupiter at opposition.

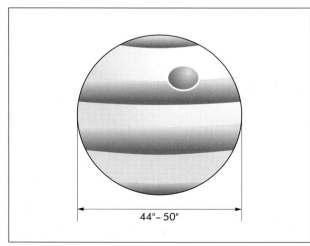

44"– 50"

wavelengths and/or larger apertures mean better resolving power, whether using an optical or radio telescope.

The *apparent field of view* (compare the last of the formulae above) is the field experienced by an observer when looking through an eyepiece. "Ultra-wide field" eyepieces can have apparent fields of 80 degrees or more. The edges of the field are not particularly noticeable and the observer seems to be floating in space rather than peering through a port-hole.

The *real field* is the field of view on the sky – the actual amount of sky that is visible. It depends on the magnification used. An 80 degree apparent field eyepiece which gives a magnification of 160× is only observing a real field on the sky of 0.5 degrees, i.e. the diameter of the full Moon.

Finally, the term *exit pupil* is often used. This is simply the imaginary disk (or *Ramsden disk*) where the bundle of light rays leaving the eyepiece is at its smallest diameter – which is, of course, the optimum position for the pupil of the eye to be positioned (it is where most observers will naturally position their eye to look through the eyepiece).

Eyepiece Sizes

This is a good point at which to mention *eyepiece sizes*, something which occasionally leads to confusion. Eyepiece *focal lengths* and *diameters* are generally specified in millimetres, but the two terms should not be mixed up!

There are three *diameters* of eyepiece in regular use:

- 24.5 mm (1 inch) – this is a very old eyepiece standard, but it is still found on the cheapest telescopes.
- 31.7 mm (1.25 inch) – this is the standard eyepiece diameter; plenty of manufacturers make them, and there are many ranges to choose from.
- 50.8 mm (2-inch) – monster eyepieces, good for delivering the widest fields and the lowest powers.

Now that we have discussed a few fundamental points (and a few misconceptions), let's move on to a detailed look at refractors.

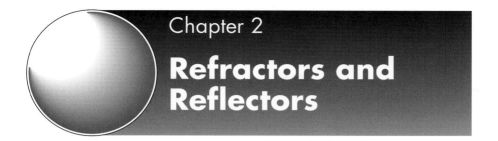

Chapter 2

Refractors and Reflectors

Achromatic Refractors

Refracting telescopes ("refractors") focus light by using a lens (or "object glass", or "objective") as shown in Figure 2.1.

A simple convex telescope lens bends and disperses light in a similar manner to a prism, causing different colours to come to a focus in slightly different planes (chromatic aberration). Not surprisingly, single lenses are avoided in all but the cheapest of "toy" telescopes. Stars seen through such telescopes appear to be surrounded by haloes of colour (see Figure 2.2, *overleaf*).

Up until the early 1980's most commercial refractors used a two-element *achromatic* (it means "without colour") lens, consisting of two types of glass. This to some extent overcomes the dispersion problem by focusing two colours (green and red, usually) in the same plane.

Figure 2.1. The refractor.

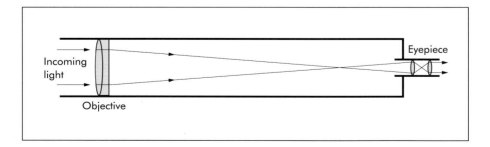

Incoming light

Objective

Eyepiece

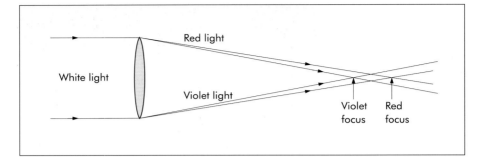

Figure 2.2.
Chromatic aberration
in a single lens
refractor.

A bi-convex (crown glass) lens and a plano concave (flint glass) lens were the standard combination. The two types of glass have different refractive indices and bend light to a slightly different extent. The addition of the second, concave, element serves to recombine the dispersed colours and minimise visible haloes of colour.

However, as you might expect, this solution is not a perfect one, and achromatic refractors always show coloured fringes around the brightest stars because they do not focus *all* wavelengths in the same plane. The problem is that an achromatic doublet can only be designed such that light of two specific wavelengths focus at the same point. Unfortunately this colour problem gets markedly worse with increasing aperture and decreasing f/ratio, so large achromatic refractors need to have very long focal ratios to avoid objectionable colour aberrations.

A rule-of-thumb is that the f/ratio of an achromatic refractor should be at least equal to the telescope aperture in centimetres to avoid objectionable colour aberrations. Applying this rule, a 10 cm refractor of f/10 has a 1 metre focal length, which is quite manageable. Unfortunately a 15 cm f/15 refractor is beginning to get unwieldy and a 20 cm f/20 refractor of 4 metres focal length is more than a handful!

In any case – and probably fortunately – achromatic refractors above 15 cm aperture are generally beyond the financial range of most amateurs, so the problem of coping with a 4-metre tube length does not arise. For refractor enthusiasts, a 15 cm achromatic refractor was long regarded as the ultimate planetary instrument (certainly, such an instrument was the pride and joy of the 1930s and 1940s stage and screen comedian Will Hay, who much preferred his 15 cm refractor over his 310 mm reflector).

Apochromatic Refractors

Since the late 1970s, a revolution in refractor design has taken place. This has come about because of three factors:

1. Advances in the fabrication of exotic forms of glass enabling hitherto unavailable "abnormal dispersion" glasses to be considered.
2. Advances in the power of computer simulation of optical systems, enabling better optical designs when using affordable glass technology.
3. The foresight of a few entrepreneurs who took advantage of the two points above and realised that there were enough amateurs around who were prepared to pay for the "perfect' refractor.

The new generation of refractors are known as *apochromats* or *semi-apochromats* and feature a much improved colour correction when compared with the standard achromatic doublet, a direct consequence of the new glass technologies which evolved in the late 1970s. The term apochromat was actually coined by Ernst Abbe in the 1870s. Abbe was one of the first optical workers to use calcium fluorite lenses. Abbe's criterion was that an apochromat would bring light of *three* different wavelengths, across the visual spectrum, to a common focus. Although apochromats did exist well before the 1970s, they were too expensive to justify mass-production.

An additional advantage that the new glass technology has brought with it is the ability to produce very short focus refractors (down to f/5) with acceptable colour aberrations. In practice, the colour correction of an apochromat is truly impressive when compared side-by-side with an achromat of the same aperture. Even extremely bright stars like Vega show only the merest suggestion of spurious colour. In fact, no lens can focus all wavelengths to the same point, apochromats simply push the majority of aberrations outside the visual range (see Figure 2.3, *overleaf*).

Needless to say, apochromats come at a price. However, hundreds of amateurs (especially in the USA) are prepared to pay that price for the exquisite images that apochromats deliver. Takahashi, in 1977, were the first manufacturer to make apochromatic telescopes using a calcium fluorite element in the

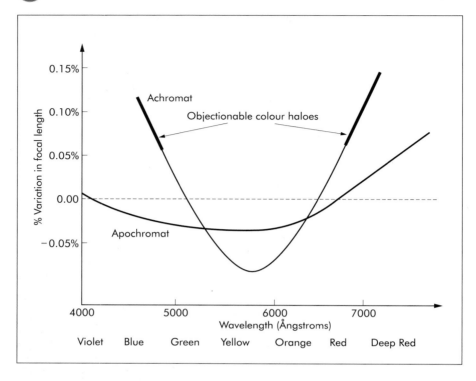

objective. Their first fluorite apochromat refractor featured a 90 mm f/13.3 objective. Since then they have offered a complete range of fluorite apochromats and the "FS" series incorporates a doublet objective with, for the first time, the fluorite element at the front of the objective to maximise light transmission and colour correction (Figure 2.4, *opposite*).

Takahashi currently (1998) offer f/8 "FS" apochromats in apertures of 78, 102 and 128 mm. An optional *telecompressor* brings the f-ratio down to just under f/6, for photographic applications. A telecompressor is simply a special lens which reduces the apparent focal length of a telescope: telecompressors are also known as *focal reducers*. Most telecompressors also flatten the field, that is they improve the quality of the star images at the edge of the field.

In 1995 I travelled to India to view the October 24th Total Solar Eclipse from the ancient Moghul city of Fatehpur Sikri. A large contingent of Japanese observers were also there and an extraordinary number of them were using Takahashi FS78 refractors on the Takahashi EM-2 mounting. Virtually alone amongst today's manufacturers, Takahashi continue to use calcium fluorite

Figure 2.3.
Variation in focal length for a 125 mm f/8 achromat and apochromat. Note how the apochromat only deviates greatly at the limits of human vision.

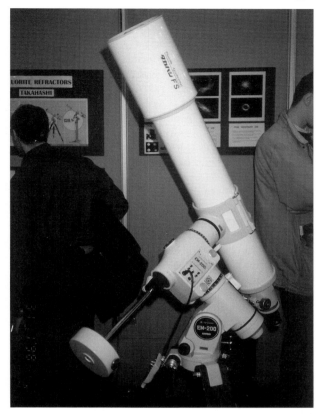

Figure 2.4. A Takahashi "FS" series 125 mm apochromat.

objectives to ensure the best possible colour correction; but it is at a price.

In the February 1963 issue of *Applied Optics* James G. Baker wrote a classic paper entitled "Planetary Telescopes". In this paper, he defined a strict criterion for the term "apochromat", namely that the optical path difference between 4047 Å (violet) and 7065 Å (red) should be less than 0.5λ (i.e. half the wavelength of light). This is a very strict criterion for a refractor to meet, and only apochromats using the most expensive glasses can hope to achieve it.

Semi-Apochromatic Refractors

In the early 1980s a number of companies realised that "affordable" designs for refractor lenses, using unusual but relatively inexpensive glasses, could yield significant improvements in performance without necessarily satisfying the Baker criterion. Strictly speaking these instruments should be termed *semi-apochromats*, although apochromat is the widely used term for them. Because the human eye is relatively insensitive in the deep-violet and far-red ends of the spectrum, the visual observer will seldom be able to distinguish between an apochromat and a semi-apochromat.

However, photographic film is a tougher test, especially at the violet end of the spectrum. In 1981 Roland Christen (*Sky & Telescope*, Oct. 1981, p. 377) exhibited a 12.5 cm semi-apochromat at the Riverside Telescope Makers convention which, by all accounts, showed far more planetary detail than any other telescope on display. This lens was of triplet design (two crown elements surrounding a flint glass central element) and made of relatively inexpensive glasses (at least, inexpensive compared with calcium fluoride).

Christen also detailed an alternative approach, the Tri-Space Refractor, in the October 1995 *Sky & Telescope*. In this design a conventional achromatic doublet passes light through a small triplet lens placed inside the focal plane. The small triplet features an "abnormal dispersion" element at its centre – again, a cheaper alternative to calcium fluoride.

During the early 1980s Christen formed "Astrophysics", a company now renowned for their high-quality apochromats which, these days, do use some fairly costly "extra low dispersion" glasses. Some of the shorter semi-apochromatic refractors (such as Tele Vue's Genesis SDF) feature an additional telecompressor/corrector lens to achieve their performance, but are no less impressive for that.

Refractors or Reflectors?

Achromatic and apochromatic refractors are the favourite choice for many planetary observers as they have a number of desirable features, namely:

● The lack of the central obstruction found in most reflectors, increasing contrast at the diffraction limit.

- A closed tube, eliminating troublesome tube currents, i.e. warm air trapped in the tube and spiralling up to the open end (see below).
- Rigidly mounted optics, removing the need for regular collimation, i.e. optical alignment (see below), checks.
- Much reduced scattered light when compared with reflective optics.

However, refractors also have a number of disadvantages, specifically:

- Achromatic refractors are expensive per inch of aperture.
- Apochromatic refractors are extremely expensive per inch of aperture.
- Achromatic refractors are less portable than Newtonians and Schmidt–Cassegrain reflectors of the same aperture.

In theory there is no reason why a long-focus Newtonian reflector with a small secondary mirror and good optics should not perform as well as an apochromat of the same aperture, and for a fraction of the cost. However, in side-by-side tests apochromats always seem to win, especially for the planets; the closed tube and rigidly mounted optics of the apochromat may be factors here. An out-of-collimation short-focus reflector with tube currents will not stand up to this sort of comparison!

In passing I want to emphasise that apochromats shouldn't be overlooked by the visual Deep Sky observer, despite their obvious advantages for lunar and planetary observers. The increased contrast available to the apochromat user can certainly pay dividends when observing the brighter Deep Sky objects. Also, some of the best Deep Sky photographs – such as those adorning the "Gallery" pages of *Sky & Telescope*, are taken by owners of six- and seven-inch apochromats. Admittedly, these photographs are always taken by the very best astrophotographers, who seem to be able to secure good images whatever equipment they use. Nevertheless, in recent years, the best Deep Sky astrophotographers have shown a marked leaning towards short focus (typically f/7) apochromats, rather than Newtonians or other forms of *astrograph* (an astrograph is a term meaning "astronomical camera").

Don't fall into the trap of buying a telescope in the belief that the telescope *alone* will guarantee success in astrophotography. All the best Deep Sky and Comet photographs are taken from very dark sites, and usually at high altitude. Similarly, nearly all the best planetary photographs are taken from sites renowned for exceptional atmospheric stability, or by observers who have gone to extraordinary lengths to obtain one good image from hundreds or thousands of planetary images.

Newtonian Reflectors

Ever since Isaac Newton constructed his small reflecting telescope in 1668–69 the Newtonian reflector has dominated the world of astronomy, especially where a large aperture is required. Strangely, despite the dominance of reflectors in professional and amateur astronomy, the average non-astronomer (at least those that I know) seem to think that a "lensless" telescope is, in some ways, inferior.

The main function of any optical astronomical telescope is, of course, to precisely focus parallel light rays, and whether this is achieved with a mirror or a lens is irrelevant. The *accuracy* of the optical surface is the important factor, and both mirrors and lenses can be fabricated to an accuracy exceeding one-quarter of the wavelength of light – the critical value for diffraction-limited resolution.

Unlike refractors, where modern techniques have revolutionised the reduction of chromatic aberration, the design of the Newtonian reflector has remained essentially unchanged for four hundred years (see Figure 2.5).

Figure 2.5. The Newtonian reflector.

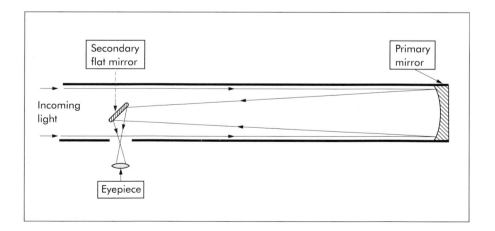

But there *has* been a dramatic change in the size of amateurs' Newtonian telescopes over the last twenty-five years. In the 1970s a 30 cm Newtonian (Figure 2.6) was widely regarded as the largest aperture an amateur might desire. As we approach the twenty-first century this anticipation has almost doubled; in the USA 60 cm optics are widely available and amateur instruments of 70 and 80 cm aperture are frequently displayed at the major star parties. I myself own a 0.49 m Newtonian – see Figure 2.7 (*overleaf*).

Although small-aperture apochromats invariably out-perform reflectors of the same aperture in side-by-side comparisons they simply cannot out-perform high-quality reflectors that have significantly more

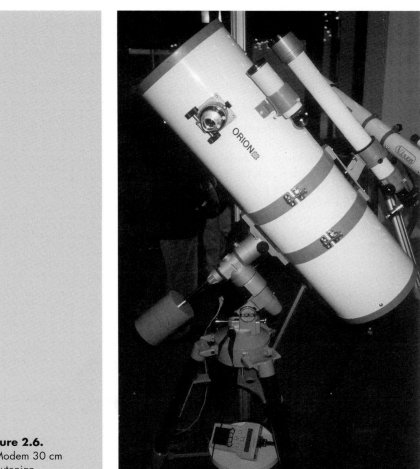

Figure 2.6.
A Modem 30 cm Newtonian.

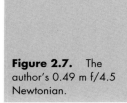

Figure 2.7. The author's 0.49 m f/4.5 Newtonian.

aperture. A 15 cm apochromat might well give a short-focus 20 cm Newtonian a stiff challenge (particularly when observing the Moon and planets) but it should not *significantly* out-perform the reflector – if it does, there is almost certainly a problem with the reflector. This can be due to inferior optics (main and secondary mirrors), eyepieces that do not perform at fast f/ratios, poor collimation, tube currents, or heat from the observer's body. The same necessity for high-quality of manufacture applies to reflectors as well as refractors.

If you want a Newtonian reflector to deliver refractor-like results make sure:

- The f/ratio is 6 or higher.
- The optics and eyepieces are first class.
- The diameter of the secondary mirror is less than a fifth of the primary; preferably much less.
- The secondary mirror support (spider) is well de-signed and the support vanes are thin.
- The optics are well collimated (that is, accurately and correctly aligned) and well supported.
- The tube is free from stray light but is well ventilated; heat from the observer's body should be prevented from drifting across the light path ("tube currents").

• The mirror is not mounted close to ground level, where turbulent heat rising from the ground is at its most destructive, i.e. it degrades the fine detail.

The larger the aperture and the thinner the mirror, the greater the danger of the mirror sagging under its own weight. As a rough guide, a 15 cm aperture mirror with a thickness:diameter ratio of 1:6 can be safely supported at three points and have negligible sag. Thinner and/or larger mirrors for amateurs should use at least a nine-point "flotation" system as shown in Figure 2.8. Mirrors larger than 45 cm aperture and/or thinner than 1:6 may well require an 18-point support system.

Beginners often query why the mirror cannot simply rest on a flat wooden or metal plate. The answer is that both the plate and the back surface of the mirror are unlikely to be perfectly flat. Thus, in practice, a mirror sitting on a flat surface will actually be supported at a handful of arbitrary points, as opposed to three, or nine precise points. A flat block of wood, or even polythene bubble-wrap (!) is OK for a small-aperture Dobsonian, but for planetary observing the mirror must *not* sag, even by a fraction of the wavelength of light.

If Newtonian reflectors are capable of delivering good results why are they so often considered to be

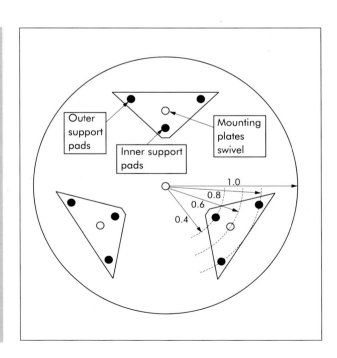

Figure 2.8.
A nine-point suspension system – three swivelling triangular plates each containing three support pads.

inherently inferior, especially as planetary instruments? Undoubtedly, *seeing* effects are a major factor here, especially those that are due to the instability of the atmosphere and not local to the instrument and observer. For planetary observers, particularly those living in a turbulent air stream, large apertures will rarely offer a significant advantage.

During the night, heat rises from the ground, something that invariably ruins fine planetary detail. In the UK (latitude 50° and higher) it is rarely possible to glimpse details smaller than 2″ (2 arc-seconds) except on nights of very good stability. Observing the planets at high magnification is often like looking through a bowl of water: the slightest turbulence will ruin the definition. The superb drawings made by the best planetary observers are built up from, maybe, ten minutes of intense concentration with the eye scouring the planet for fleeting moments of good seeing. But even the best moments rarely allow details smaller than 1″ to be seen.

When you consider that a 10 cm aperture instrument (see Figure 2.9, *opposite*) can theoretically resolve 1″ it follows that small apertures can often deliver as much resolution as large apertures under typical observing conditions. However, inexperienced observers will almost certainly prefer the larger aperture option: planetary details at 200× can appear rather faint to the untrained eye.

Another factor perpetuating "the refractor myth" is that refractors of 10–20 cm aperture are rarely made by amateurs; they are usually acquired commercially from quality telescope manufacturers. Conversely, many Newtonians are home made and while the primary optics may sometimes be excellent, other design features (see above) are often neglected.

The trend, in the last twenty years, has been to build larger and larger Newtonians of fast f-ratios: typically f/4.5, but sometimes f/4.0, or even f/3.5. Modest instruments of this speed are sometimes called "Richest Field Telescopes" as their aperture and wide fields of view can show an enormous number of stars at low powers. With reflectors, any faster f-ratios and the secondary mirrors compromise both light grasp and field of view, as well as suffering from severe optical aberrations (indeed, Newtonians faster than about f/6.0 start to suffer noticeably from an aberration called *coma*, where the stars turn into "seagull" shapes away from the telescope's optical axis).

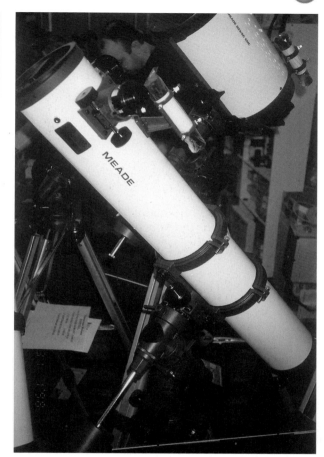

Figure 2.9. An ideal telescope for beginners, a 10 cm Newtonian.

The US telescope manufacturer, Parks, make a standard range of Cassegrain–Newtonians (dual focal length instruments) with a Newtonian f-ratio of f/3.5! This drift towards larger and faster Newtonians has come about because of a number of factors:

- Modern glass technology enabling large, but thin, lightweight mirrors to be fabricated.

- A desire by amateurs to get ever larger optics ("aperture fever").

- The CCD revolution; being smaller and of higher resolution than photographic film, CCDs (electronic imagers) need shorter focal lengths for an optimum image scale (typically 2″/pixel) and sensible field of view.

In this mad rush to collect more photons, and image objects of magnitude 20 and beyond, one type of telescope, once the mainstay of the amateur planetary observer, seems to have been neglected. This is unfortunate, as many experienced observers regard the *long-focus Newtonian* as, perhaps, the ultimate amateur telescope.

Long-Focus Newtonians

Ever since the first editions of *Amateur Telescope Making* appeared in the late 1920s, amateur astronomers have been advised of the virtues of the long-focus Newtonian reflector (see Figure 2.10).

Franklin Wright, in the first volume, pointed out that telescope mirrors of f/8 or slower were far easier to

Figure 2.10. A simple 22 cm f/8 Newtonian, ideal for planetary observing.

figure, to deliver diffraction-limited performance, than shorter-focus instruments. To understand the reasons for this we need to digress (briefly) into the "black art" of mirror making. Grinding and polishing telescope mirrors is nonetheless an art which many advanced amateurs have mastered.

These days more and more amateurs are now buying, rather than building, their own optics although the specialised subject of mirror making is beyond the scope of this equipment guide. Nevertheless, the subject is covered by a number of excellent books on the subject (see Appendix 1).

In making a Newtonian mirror, the aim is to grind and polish the mirror to produce a paraboloidal curve. Only a parabola will focus parallel light rays to precisely the same focal point. After initial grinding and polishing, the mirror will usually end up with a spherical surface; parabolising by further polishing is then required. A short-focus mirror (a mirror having a low f-ratio) has a deep curve, and there is an appreciable difference between the spherical and paraboloidal curves; the mirror must be parabolised or the performance will be unacceptable.

However, a *long*-focus mirror is ground to a shallow curve, and the difference between the spherical and paraboloidal surface is less significant. Indeed, at around f/12 the difference between the paraboloid and the sphere is negligible and parabolising the mirror may well be unnecessary. Put simply, a long-focus Newtonian mirror is less critical to figure than one of short focus, and there is more chance of producing a superb mirror – with a surface figure close to perfection – when the mirror has a large f-ratio.

An additional benefit of a long f-ratio mirror, intended for high-resolution, narrow-field work, is that the secondary mirror can be significantly smaller (Figure 2.11, *overleaf*). This is a consequence of the primary-to-secondary distance being, perhaps, ten times greater than the secondary-to-eyepiece distance. Reducing the size of the secondary mirror is not simply to increase the light transmission. The presence of a large secondary mirror produces undesirable *diffraction* effects that reduce contrast in the subtle detail at the telescope's limit, making (for example) fine detail in Jupiter's belts and zones harder to see. The late Horace Dall, the renowned English telescope maker, carried out a series of tests in which he increased the size of a telescope's central obstruction and noted the effect on resolution at

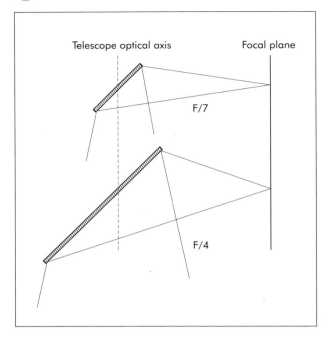

Figure 2.11. A fast Newtonian needs a large secondary mirror. In addition the mirror should be placed asymmetrically in the tube as the light cone is smaller at the top edge of the secondary.

the telescope's limit. He concluded that the diameter of the secondary obstruction should be less than 20% of the primary diameter to ensure no ill effects.

Visual planetary observers are frequently more critical than this, often recommending secondary diameters smaller than 12% to 15% of the primary diameter. Either way, a long-focus Newtonian with a small secondary mirror could well be the ultimate planetary instrument.

Another benefit, often overlooked, is that long-focal-ratio Newtonians are significantly easier to collimate than their shorter-focus counterparts (see Figure 2.12, *opposite*).

Collimation

Collimating a telescope is simply the act of aligning all the mirrors for optimum performance. Refractors and Maksutovs (see below) come factory-set, but Newtonians and Schmidt–Cassegrains (also see below) permit – and sometimes require – some adjustment.

For Newtonian telescopes there are two stages to the collimation process: firstly, collimating the optics in daylight, by eye; secondly, fine-tuning the optics on a

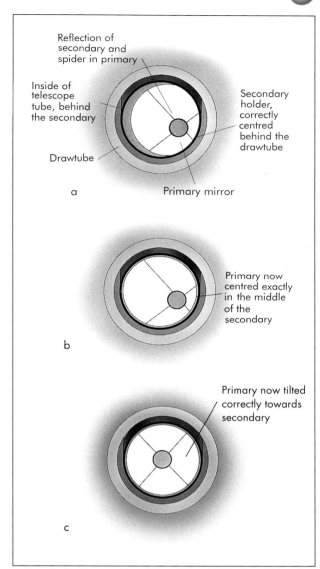

Reflection of secondary and spider in primary

Inside of telescope tube, behind the secondary

Secondary holder, correctly centred behind the drawtube

Drawtube

a

Primary mirror

Primary now centred exactly in the middle of the secondary

b

Primary now tilted correctly towards secondary

c

Figure 2.12. Three steps in collimating a Newtonian reflector – this is far less critical in a slow Newtonian.

star. For both operations it is one hundred times easier if two people are involved instead of just one! Figure 2.12 shows the basic steps for collimating a Newtonian reflector in daylight. A 35 mm plastic film canister with a hole punched in the base makes a good sighting tube. Most telescope manuals will supply full details for each specific model. Note that in a fast, f/5 or f/4, Newtonian the flat is not positioned symmetrically in the tube; Figure 2.11 illustrates this point well.

Once daylight collimation has been finished, the mirrors can be "tweaked" such that the out-of-focus diffraction rings surrounding a bright star appear concentric. An alternative, easier, but more expensive approach is to use a laser-collimating tool.

To go back to my original point: it's easier to collimate a long-focus Newtonian.

Many leading planetary observers extol the virtues of long-focus Newtonians. The late Clyde Tombaugh, discoverer of the Planet Pluto, built his first long-focus Newtonian (a 22.8 cm f/8.8 instrument) in 1928. His ultimate instrument, a 40 cm f/10 monster, enabled the "spokes" in Saturn's rings to be glimpsed years before they were imaged by the Voyager spacecraft.

The veteran American planetary observer, Thomas Cave, built his first long-focus Newtonian, a 15 cm f/10 instrument, in 1935. As described in the British Astronomical Association's *'Instruments & Imaging' News* (Vol. 1, Issue 2, 1994) he then progressed to a 20 cm f/10, a 25 cm f/11.2 and, finally, a 32 cm f/10.7 which "was the very ultimate in perfection ... the ultra fine planetary details were amazing".

The obvious question is, if long-focus Newtonians are the ultimate planetary instrument (and *much* cheaper than the equivalent aperture refractor!), why are they rarely possessed by amateurs? The main reason is the sheer length of the optical assembly and the corresponding tube and mount size. Three or four metre long tubes are impractical for mass manufacture and require substantial observatories (not to mention observing ladders!).

And manufacturing and testing the mirror are not entirely plain sailing: although the precise figuring of the optical surface of the long-focus mirror is less critical, testing the mirror figure with the standard Foucault and Ronchi shadow patterns is quite a challenge; the shadows are much fainter than those seen with mirrors of smaller f-ratio. Perhaps this is why long-focus Newtonians of significant aperture will probably remain the possession of a few dedicated amateurs.

However, smaller-aperture Newtonians are typically manufactured at f/ratios up to 8 and these relatively small instruments can deliver excellent planetary views.

Large-aperture Newtonian telescopes are typically sold as fast 'Dobsonians'. These instruments may not deliver the ultimate planetary views, but they are extremely user friendly and great fun to use; so let us now have a close look at the Dobsonian reflector.

Dobsonians

The Dobsonian reflector is named after John Dobson, an ex-monk and amateur astronomer who has spent most of his life promoting the joys of casual visual observing. A Dobsonian is, in fact, essentially no more than an alt–azimuth-mounted Newtonian reflector – so what's all the fuss about?

Before the arrival of the commercial Dobsonian reflector, mass-produced Newtonian telescopes were either very small alt–azimuth instruments or larger, equatorially mounted instruments. The thinking behind this was that large telescopes needed to be mounted on an equatorial mounting, with either slow motions or an electric drive to counter the Earth's rotation. After all, large telescopes use large magnifications and you can't comfortably use a large telescope unless it is equatorially mounted with a drive – right? Well, not really!

Firstly, there are many amateur astronomers who enjoy scanning the wide-field show-pieces of the sky (the Orion Nebula, Andromeda Galaxy, Double Cluster etc.) at low powers, and the minor inconvenience of nudging the telescope tube every thirty seconds or so, to compensate for the Earth's rotation, is relatively trivial.

In fact, my friend Richard McKim, the renowned British planetary observer, uses high powers on his 30 cm f/7 alt–azimuth-mounted Newtonian and claims that the drift of a planet through the field actually enhances planetary detail.

This is by no means far-fetched. The eye–brain combination does not perform well when staring at a fixed object. As anyone who has tried guiding on a bright guide star will know, after staring for a while, the star seems to disappear; but look away for a moment and the star reappears. Also, equatorial mountings can be a real hassle for the visual observer: apart from the set-up time for polar alignment, the eyepiece always seems to end up in the most inconvenient position!

The altitude and azimuth bearings of the Dobsonian reflector are mounted at a very low level which is very important for maximising observing comfort (see Figure 2.13, *overleaf*).

The azimuth bearing is in fact only just above ground level and, because almost all of the telescope's weight is in the primary mirror, the altitude bearing/ balance point can be positioned at knee-height in all but the largest Dobsonians. The low height of the bearings and

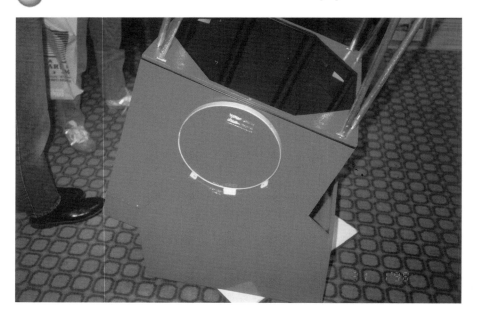

the fact that they move in altitude and azimuth make life relatively comfortable for the visual observer, who has only to move around the telescope in a circle. There is no climbing of ladders in the dark – at least when using modest apertures – as you might well need to do when using an equatorially mounted Newtonian on a plinth. Objects at low altitude can be catered for by sitting on a chair or an observing stool.

The bearings of a Dobsonian are usually made from low-friction materials such as Teflon (polytetrafluoro-ethylene); this provides an ultra-smooth movement with the least effort. In the 1980s inexpensive techno-logy became available which enabled *digital setting circles* (i.e. electronic equivalents to the engraved posi-tion circles often fixed to telescope mounting shafts) to be attached to alt–azimuth Dobsonians. By solving spherical co-ordinate geometry equations with micro-electronics, astronomical equipment manufacturers could then attach shaft encoders and associated gad-getry to alt–azimuth telescopes, enabling the observer to know where he was pointing the telescope, despite not using an equatorial mounting.

These devices can also help guide the observer to thousands of Deep Sky objects as well as displaying the Right Ascension (Astronomical East/West) and Declination (Astronomical North/South) of the observed field.

Figure 2.13. The huge Dec bearings of Richard Fleet's 50 cm Dobsonian.

An increasing number of Schmidt–Cassegrain telescopes are now sold on alt–azimuth mountings that are able to track in an equatorial manner. We will take a look at these very popular instruments in the next chapter.

Buying a Telescope

Having looked at the arguments for and against refractors and reflectors, the question arises of who to buy the instrument from.

As a British observer, it is rather painful to admit that the thriving British telescope industry of the 1960s and 1970s is no more. Most of the British companies that made telescopes in those times have now either disappeared or moved into telescope importing. The vast majority of telescopes purchased in the UK are imported from the USA or the Far East, particularly Japan. Although some UK-made telescopes (mainly Newtonians) are still produced, they simply don't have the features available on imported telescopes.

In recent years, the US company Meade has dominated the world telescope market by investing in new technology and by listening to customer feedback from its clients. The previous "brand leader" in this market – Celestron – still survives and still produces high-quality equipment, but at the moment Meade always seems to me to be one step ahead.

As far as refractors are concerned, there are a number of relatively small manufacturers who co-exist with the giants and who (in the main) have survived by producing products at the very top of the range, for example very high-quality apochromats.

Beginners' Telescopes

For some reason many children go through a stage, at the age of eleven or so, where they are fascinated by astronomy and want to own a telescope. In most cases the interest is relatively short-lived and the negative aspects of the hobby (i.e. venturing outside on cold, dark and damp nights and not being able to find or see what you want to) take their toll on a young person's enthusiasm. It is unusual for the interest to survive the teenage years: parties, relationships, exam pressure and a lack of finance are usually responsible. And even if the interest survives into the observer's twenties, career pressure and the responsibilities of family life can be the final death-knell.

In my own case, shortly after my tenth birthday I picked up a friend's birthday present, which my parents had bought for him. The present was the brand new, 1968 edition, of *The Observer's Book of Astronomy*, by Patrick Moore. Before I saw this book I had just not realised that useful observations of the Moon and planets could be made by amateurs! It was the pictures of amateur telescopes in this book and the knowledge that amateurs could make useful observations that fired my particular interest in astronomy. An enthusiasm – almost an obsession – for acquiring bigger and better telescopes fuelled my interest, and a few vital observing triumphs along the way kept me going.

Good-quality equipment that is easy to use can make all the difference!

Buyer Beware!

For newcomers to astronomy there is always a temptation to buy the most attractive small telescope available in the high street. In the UK, at least, this can be a gamble. There are a number of unscrupulous manufacturers from the Far East who supply very cheap telescopes to high street chain stores, who stock them, in the main, unknowing just how bad they are. Don't be fooled, you get what you pay for. Sad but true.

Before buying a first telescope check off the following points:

- Pay a visit to your local astronomical club. Almost every town has one. Ask to have a look through some small telescopes and listen to their owner's comments.
- Contact your national astronomical society for advice.
- Read the equipment reviews in astronomy magazines.
- Check out where your national astronomical telescope dealers are based (the ads in the magazines will help). Send for their catalogues or, if they have a showroom, pay them a visit. In general, they will be able to offer far better advice and far better telescopes than your local camera shop.

• If you do decide to purchase a small refractor from the high street, be very sure about it! Don't be fooled by a glossy enamel finish. Examine the quality of workmanship before you purchase and check that the refractor's lens is not "stopped down" by a metal ring, just behind the lens – a sure sign of poor optics! In particular, check that the mounting – usually a tripod – is stable.

There isn't space here to describe every manufacturer's product line but a brief mention of some suppliers of good-quality instruments might be helpful.

Suppliers of good refractors include Celestron, Meade, Astrophysics, Tele Vue and Takahashi, and these are mentioned elsewhere in this chapter; but Orion, Vixen, Swift, Parks, Unitron, Bausch & Lomb and Edmund Scientific also supply good refractors in the 60–80 mm range (see Figure 2.14).

There is not much to choose between the best of the smallest refractors, and your choice may depend on whether or not a telescope dealer near you stocks a particular brand. However, if you choose a small refractor from the reputable companies detailed in this chapter, you should (unless you are extremely unlucky) end up with a good instrument.

For many of us, in the 1970s and 1980s a Celestron was the telescope to own. Their affordable, mass-produced, 20 cm Schmidt–Cassegrains were the ultimate portable instrument and they led the way for twenty years. But inexorably, the Meade Corporation went from strength to strength, capturing more and more of Celestron's market and constantly improving

Figure 2.14.
A small-aperture (80 mm) refractor from Vixen.

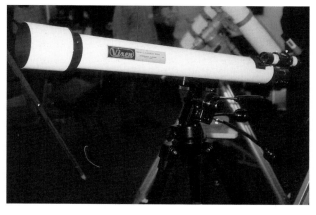

its own products. Celestron have survived though, and, by diversifying (e.g. to German Equatorial mountings – see Chapter 6 – for their Schmidt–Cassegrains), has managed to maintain a loyal following.

The Celestron Firstscope 60 mm and 80 mm models are excellent beginners' telescopes. Just looking at them brings back memories of the excitement I felt when I purchased my own 60 mm refractor after a year of saving my pocket money! At last I could see the rings of Saturn and the moons of Jupiter, something I could never do with my low-power 30 mm refractor. That first look through my new refractor was only bettered by my views, as a teenager, through the 100 mm refractor, mounted on the roof of the Bury St Edmunds Athenaeum! No other views I have had since, however expensive the equipment, have been as exciting; and I have heard similar stories from other observers, not least the late Harold Ridley, a lifelong British meteor and comet photographer. Maybe those first experiences, at such an impressionable age, simply can't be matched?

I have heard the view expressed, by a number of veteran observers, that a refractor below 75 mm (3-inch) aperture is just too small to be of any use; if an observer's budget imposes a restriction to a 60 mm refractor, a pair of binoculars would be better. The temperament of the observer is the vital factor. Certainly, as a schoolboy, my overwhelming desire was to buy a telescope that could show me lunar craters and detail on the planets. I wanted something that could provide MAGNIFICATION!

These days I would certainly derive more observing pleasure from using a pair of low-power, wide-field binoculars, but I very much doubt whether a new pair of binoculars would have fired my enthusiasm at the age of 13, to the same degree as a shiny new refractor!

But if a 60 mm refractor is too small to be of use (which I would dispute) an 80 mm refractor is certainly worth aiming for. In the 1970s an 80 mm refractor would have been a schoolboy's dream telescope, and I suspect this is still the case. At this aperture, a number of companies offer a German Equatorial mounting with an electric drive; needless to say this adds considerably to the cost of a small instrument.

It is also important to remember that, at these small apertures, chromatic aberration is *not* a problem in a quality instrument of f/11.

High-Quality Refractors

The top refractor in Celestron's line is the semi-apochromatic 102 mm aperture f/9 GP-C102ED.

This model, like its achromatic cousin, is distinguished by a sleek, black tube and the excellent and reliable GP ("Great Polaris") mount. The "Great-Polaris" mount is actually manufactured by the Japanese telescope company Vixen and is renowned for its reliability at an affordable cost (see Figure 2.15).

During the close approach of Comet Hyakutake in 1996, I travelled, with two colleagues, to Tenerife and a Vixen GP mount served as our main camera platform for unguided exposures with 50 and 85 mm lenses. The mount survived the trip, was easy to polar align and operate and tracked perfectly (at those focal lengths) for a hundred exposures over four nights.

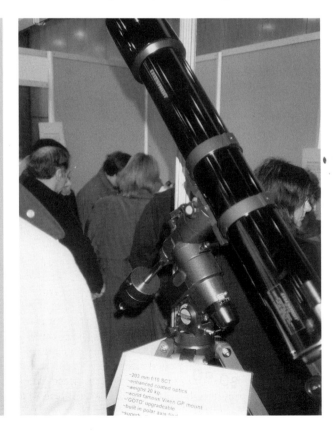

Figure 2.15.
A Celestron 102 mm refractor on the Vixen "Great-Polaris" mount.

Astronomical Equipment for Amateurs

Celestron's top refractor won't whiz around (slew) and locate objects in the sky (the equivalent Meade will), but optional "Advanced Astro Master" digital setting circles can help you around the sky without too much trouble.

The "giant" of the telescope market, Meade, also produces a comprehensive line of refractors, both achromatic and semi-apochromatic, from the beginners' 60 mm refractor to the larger 102 mm to 178 mm f/9 semi-apochromats. Meade's semi-apochromats come with a German Equatorial head (LXD650 or 750) and optional #1697 Computer Drive system with (assuming precise polar alignment and calibration) arc-minute pointing precision to any RA and Dec. or any object in the 64,000 object database. In addition, the mounting can accept control signals from an *autoguider* (see Chapter 8) and features permanent *periodic error correction* to enable the user to train out any idiosyncrasies of the telescope's worm drive (see Figure 2.16)

Only Meade, Astrophysics, Zeiss and Takahashi have made apochromats of 150 mm aperture and larger in recent years. The Meade f/9 semi-apochromat lenses are of a doublet design utilising Extra Low Dispersion (ED) glass whereas the Astrophysics and

Figure 2.16.
A Meade 178 mm f/9 refractor on the LXD750 mount.

Zeiss (and the older, large Takahashi) apochromats feature a triplet lens (ED glass in the Astrophysics instruments, fluorite in the Takahashi and Zeiss instruments). An f/5.7 focal reducer/field-flattener is available for 35 mm photography.

Apochromats for the Connoisseur

Roland Christen (mentioned earlier in this section) brought Astrophysics refractors to the marketplace using the technology of Super ED glass (extra low dispersion) as the optical sandwich between two crown glass lenses. Apertures from 102 to 206 mm (!) are available.

By using the highest (and most expensive) grades of ED glass (i.e. the glass with the lowest dispersion of the optical spectrum), Christen has ensured that his apochromats' colour correction is excellent, even at relatively fast f-ratios.

Astrophysics also manufacture a fine range of German Equatorial mountings for their refractors. As experienced astrophotographers will be well aware, there are few more irritating sights than a *vignetted* astrophoto. Vignetting manifests itself as a falling off of light at the edge of the photograph and is caused by the light cone being restricted by the throat of the drawtube (the tube emerging from the focuser, into which the eyepiece/camera adapter is inserted). The larger the drawtube, the less the chance there is of vignetting, especially with medium format film. Astrophysics instruments feature enormous drawtube diameters: 70 mm in diameter for most models, but an enormous 100 mm in diameter for their two 155 mm and 206 mm EDF models, which employ massive field-flatteners at the drawtube end.

Astrophysics' 155 mm f/7.1 EDT and EDF apochromats have proved extremely popular with dedicated Deep Sky astrophotographers. In addition, Astrophysics' 105 mm f/5.8 "Traveller" apochromat has become the instrument of choice for many solar eclipse chasers who want to take an equatorially mounted astrograph to far-flung locations without exceeding their airline baggage allowance.

An alternative approach to short-focus refractor design is practised by Tele Vue. Another well-established force in the small aperture apochromat market, Tele Vue was formed in 1977 by optics expert Al Nagler, a veteran of the 1960's space program engineering effort and designer of wide-field simulators for NASA's Gemini and Apollo programs. Tele Vue started a revolution in quality eyepiece design (see Chapter 5) but soon branched out into the small refractor market.

Over the past 15 years a number of Tele Vue refractors have been aimed at the discerning casual observer, but none has been more attractive than the Genesis SDF, a 102 mm f/5.4 system with excellent colour correction – remarkable at such a short f/ratio (shown in Figure 2.17). The unusual optical configuration consists of a two-element f/12 front objective and a two-element corrector lens/telecompressor placed before the focal plane. The telecompressor brings the final f/ratio down to 5.4. Nagler claims that by fine-tuning the air gap in both sets of doublets he can optimise the performance of each telescope.

The objective features "special dispersion" (SD) elements (possibly ED glass, but Tele Vue don't say) and the corrector/telecompressor incorporates a calcium fluorite (F) element. The relatively small size of the second doublet minimises the cost of the fluorite component.

The Genesis SDF can also be purchased as part of the "Sky Tour Observing System", a $2000 system including

Figure 2.17. The TeleVue Genesis SDF 102 mm f/5.4.

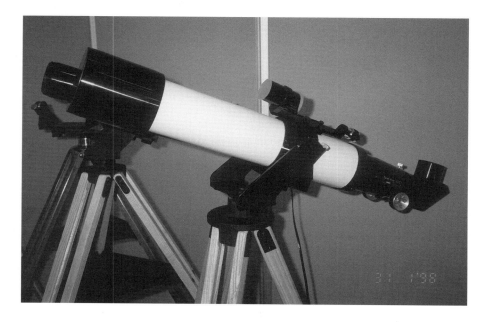

an observing stool and "Sky Tour" computer with a 2,000 object database in a calculator-sized unit. If your finances don't stretch to $2000, Tele Vue make the "Pronto" and "Ranger" 70 mm aperture f/6.8 refractors to suit a smaller budget. These models are suitable for astronomy or bird-watching and are highly portable optical assemblies.

Other Telescope Considerations

In this chapter we have looked at refractors large and small, from specialist instruments for the connoisseur, to instruments for the beginner. The telescope manufacturers mentioned are all reputable companies and their products should not disappoint. As with the other categories of telescope discussed in this book, the final choice may be determined by the availability of telescopes in your region.

There are no major refractor manufacturers in the UK, so all the best refractors, certainly all the models described in detail above, are imported. Obviously, importing adds to the cost of a telescope; apart from the import duty and local taxes, importers have to add on their own overheads and profit margin. In this respect amateurs in the USA and Japan are considerably better off than their British colleagues; they can buy major brand telescopes which are made in their own countries.

There are, of course, ways around using importers. Small items, like eyepieces, can be purchased by mail order, or while on holiday in the USA (but you still have to pay duty and taxes). I have used both methods successfully myself.

For larger items such as complete telescopes, most UK amateurs will prefer to buy from a reputable dealer/importer. After all, if there is a defect in your brand new refractor or reflector, you won't want the hassle and expense of shipping it back; this is where local dealers can bring peace of mind and justify their mark-up.

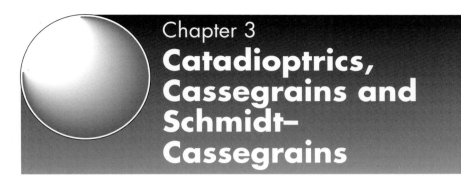

Chapter 3

Catadioptrics, Cassegrains and Schmidt–Cassegrains

Schmidt–Cassegrains are often referred to as *catadioptric telescopes*, a term applied to telescopes which use a mixture of mirrors and lenses to form the image at the focal plane.

The ubiquitous Schmidt–Cassegrain dominates the catadioptric group. This type of telescope has an appeal second only to the simple Newtonian. Despite the complex optical manufacture involved, a mass-produced 20 cm Schmidt–Cassegrain (or "SCT") on an equatorial mounting will only cost you about twice as much as the equivalent Newtonian, or around three times as much as a 20 cm Dobsonian.

Cassegrains

A standard "classical" Cassegrain telescope collects light via a mirror, just like the Newtonian. However, instead of the secondary mirror deflecting the converging light beam at right-angles through the side of the telescope tube to the eyepiece, a convex mirror (usually hyperboloidal, but spherical in the Dall–Kirkham design) sends the light beam back down the tube and out through a hole in centre of the primary mirror (Figure 3.1, *overleaf*).

The Cassegrain secondary diverges the converging rays of light somewhat. At the eyepiece the primary mirror looks as if it is further away then it is, which

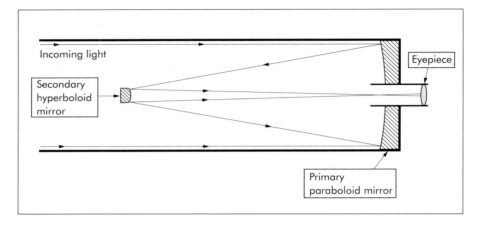

increases the focal length – typically by a factor of four. (This is similar in principle to the way a *Barlow lens* works: see Chapter 5.)

In practice, a Cassegrain has a much shorter tube length for a given effective focal length, and better eyepiece performance because of the higher f-ratio. Typical Cassegrains have primary mirror f-ratios of 4 or 5 and final (effective) f-ratios of 16, 20 or even higher.

The optical genius and British amateur astronomer, Horace Dall, used a 380 mm aperture *Dall–Kirkham* of his own design which featured extraordinarily large effective f-ratios. A modern Dall–Kirkham Cassegrain is shown in Figure 3.2 (*opposite*). Dall under-corrected the parabolic primary mirror so that a simpler – spherical – secondary mirror could be used. His primary mirror was f/5, but he used a whole range of home-made secondary mirrors and transfer lenses which produced remarkable planetary photographs at f-ratios up to f/200!

Figure 3.1. The Cassegrain reflector.

Schmidt–Cassegrains

The Schmidt–Cassegrain design is an even more effective "tube length shortener", its main advantage – and the reason for its immense popularity – being sheer compactness, as well as an eyepiece position which varies little from horizon to zenith. The observer's comfort and convenience are maximised and even large instruments (up to 40 cm) are transportable. The smaller 20 cm instruments are highly portable.

The primary mirror of commercial Schmidt–Cassegrains is typically f/2, and the secondary typically

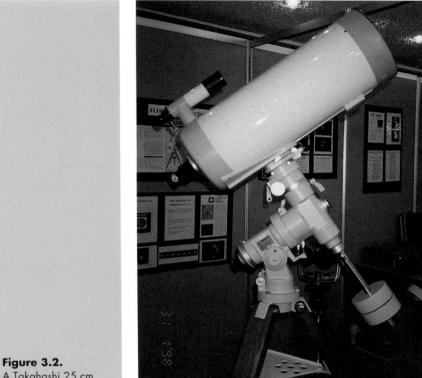

Figure 3.2.
A Takahashi 25 cm
Mewlon Dall–Kirkham.

amplifies this to f/10. Aberrations inherent in such a fast (f/2) design are reduced by the use of a two-sided (that is, figured on both sides) aspheric Schmidt corrector plate at the top of the tube. The corrector plate seals the tube, minimising tube currents and provides a convenient (and diffraction spike-free) mounting for the secondary mirror (Figure 3.3, *overleaf*).

The primary mirror of a commercial Schmidt–Cassegrain is generally 3% or 4% larger than the telescope aperture, which is determined by the diameter of the corrector plate; this guarantees that the whole of the field of view is fully illuminated. The secondary mirror is of a convex, aspheric design.

For optimum performance the optical system of a Schmidt–Cassegrain must be very well designed. In any Cassegrain system there is a risk of stray light from the sky spilling past the secondary mirror support and shining directly into the eyepiece, flooding the field of view, a risk that is especially high when the Moon is

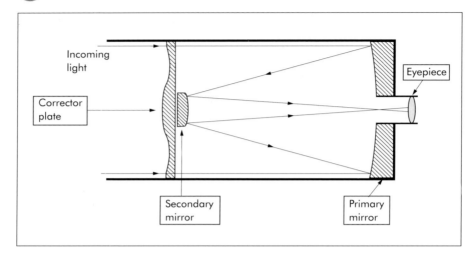

Incoming light

Corrector plate

Eyepiece

Secondary mirror

Primary mirror

being observed. Simply making the secondary mirror support very large would prevent this but would cut off more light than is desirable; it would also result in unpleasant diffraction effects which would degrade the optical performance. Primary and secondary *baffle tubes* are used to block stray light: their design is critical. Both Meade and Celestron use strategically placed *field stops* in their primary baffle tube to prevent scattered light degrading the image.

If you were to have a Schmidt–Cassegrain made as a custom "one-off" instrument it would cost at least five times the current commercial price (depending on whether or not you imported the instrument)! Mass production for the American market has brought this compact design within the financial reach of many amateurs. The two major players in the Schmidt–Cassegrain market are Celestron (since 1970) and Meade (since 1980).

Originally, in the 1970s and up to the late 1980s, Celestron were the market leaders and had pioneered the Schmidt–Cassegrain revolution. In the 1960s Celestron's Thomas J. Johnson found a way of mass-producing the complex optics and corrector plate. Here at last was an ultra-compact telescope with a tube length barely twice the mirror diameter and a range of accessories to enable long-exposure Deep Sky astrophotography.

Now (in the late 1990s) the range of accessories is staggering. From the late 1980s onwards the rivalry between Celestron and Meade intensified to the advantage of the amateur astronomer: the then-existing disadvantages and shortcomings of these telescopes were

Figure 3.3. The Schmidt–Cassegrain.

urgently addressed in an attempt to win over customers. For example, one of the problems with the original Celestron SCTs was that despite being advertised as "suitable for long exposure deep sky photography", the f/10 focal ratio meant that exposures of one or two hours were necessary to record faint extended objects such as galaxies and nebulae. Many US amateurs were forced to chill their film to −70°C or *gas-hypersensitise* it (see Chapter 7) to reduce their exposures to sensible times.

Another early difficulty was that the inherent curved focal plane and vignetting of the SCT meant that only the central 20 mm of the film was usable: a similar problem to that caused by coma, faced by owners of fast Newtonians, except that the actual field of view was much smaller.

Because of these shortcomings, a number of dedicated US Deep Sky photographers capitalised on the strongest points of the Schmidt–Cassegrain design and specialised in taking *very* long, narrow-field Deep Sky exposures of galaxies from dark, mountain-top sites (such as Mount Pinos in California). Mind you, three-hour manually guided exposures were only for *really* obsessive astrophotographers!

The work of US amateurs undoubtedly helped to promote SCTs as Deep Sky instruments in Europe and in the UK as well as in the US, despite the fact that the average UK amateur using an unmodified SCT (under traditionally hazy British skies) could never hope to match the excellence that US observers achieve.

Since 1988, Meade have offered an f/6.3 version of their 20 cm and 25 cm SCTs, which has a wider field of view, albeit at the expense of greater optical aberrations. Such aberrations preclude this option on their 30 cm and 40 cm SCTs. Both Celestron and Meade offer 0.63× telecompressors/field flatteners to bring the f/10 focal lengths down to more manageable levels for Deep Sky work. My own 30 cm Meade LX200 is shown in Figure 3.4 (*overleaf*).

Electronic Drive Control

Since the late 1980s, both Meade and Celestron have increasingly employed sophisticated electronics on their SCT ranges. Not only has this made the telescopes more user- friendly, it has also made them more "fun", so that with digital slewing and positional readout even novice observers can view dozens of objects each night.

Figure 3.4. The author's 30 cm Meade LX200.

Although electronic *positioning* and *tracking* are the most important features of these drive systems, they can also compensate for the inevitable inadequacies of the small (by professional standards) worm and wheel drive systems employed in SCTs by microprocessor wizardry.

The slewing and tracking electronics of the Meade LX200 are shown in simplified form (for one axis only) in Figure 3.5 (*opposite*). It is remarkable that such mass-produced sophistication is available to the amateur. If you take an LX200 apart (don't!), shivers go down the spine when you first see the size of the DC motors used for tracking and slewing (Figure 3.6, *opposite*) – but the 10,800:1 gearing gives a huge increase in the torque available at the shaft.

The most important design feature of the LX200 series is *affordable* technology that enables it to know precisely where it is pointed in the sky. The Meade main circuit board is shown in Figure 3.7 (*overleaf*). Because 360° shaft encoders resolving to arc-minute accuracy are prohibitively expensive, the encoders are

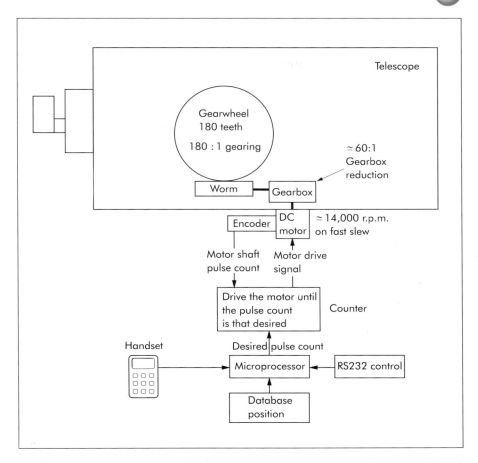

Figure 3.5. LX200 slewing and tracking block diagram.

Figure 3.6. The tiny DC motor and worm assembly of the LX200 series.

Figure 3.7. The LX200 main circuit board, housed in the drive base.

built into the motor assembly where the resolution requirement is much less critical. In fact if it were possible to build *perfect* mechanical parts, this arrangement would enable arc-second precision slewing. However, gearbox backlash, mechanical flexure and mirror movement all conspire to reduce this theoretical precision to around 5 arc-minutes or so.

Periodic Error Correction

Because telescopes need to compensate for the Earth's rotation, they have to follow the apparent movement of celestial objects at a rate of 15 arc-seconds for every second of time (for an object on the celestial equator). Such tracking needs to be precise for good photography, and a tracking error of only a few arc-seconds will show up on a photograph or CCD image as "trailed" stars.

It isn't possible to rely solely on the mechanical accuracy of the worm and wheel assembly and a fixed tracking speed, simply because the worm and wheel assembly would need to be machined to impossible accuracies to achieve that kind of precision. It is, however, fortunate that because of the way gears are machined, the deviation of a worm and wheel from "perfect" is largely repetitive – for each revolution of the worm, the tracking will deviate by the same amount at the same point in the worm cycle.

Both Meade and Celestron have used the predictability of the error (know as a *periodic* error) to make "Smart Drives" or "Periodic Error Correction" (PEC) part of their SCT drive systems. To use PEC, an observer switches the Periodic Error Correction system into "learn" mode and then manually guides the telescope with the hand controller, to keep a test star centred on the cross hairs of an illuminated reticle eyepiece. A microprocessor system linked to the telescope drive records the corrections made by the observer for one worm rotation (typically eight minutes) and then applies these corrections itself for all future rotations of the worm.

"Drive-training" can dramatically decrease the periodic worm error from, typically, 30–40″ to as low as 2–5″, a major improvement.

Autoguiders

CCD autoguiders, interfaceable to SCT drive systems, can take over where the Periodic Error Correction leaves off (i.e. at the 5″ level) and provide equally dramatic improvements in guiding accuracy. When a CCD autoguider chip "locks on" to a star – and assuming the basic tracking is fairly good, as it should be once the Periodic Error Correction is engaged – the autoguider can adjust the telescope drive such that the field is tracked to sub-pixel accuracy: as good as you need, in fact.

Once a telescope can be slewed and positioned accurately and the field can be displayed on a computer screen, observing indoors using PEC and a CCD autoguider can becomes a reality for amateurs as well as for professionals. It isn't entirely easy. As you will see in Chapter 8, CCD chips are small and so are their corresponding fields of view, so unless you can slew to an accuracy *better* than that of the CCD field of view (and

then recognise the field), remote observing is impossible. Undoubtedly, the availability of the Hubble Guide
Star Catalogue and Palomar Sky Survey on CD-ROM is
an additional help here. Even if you don't know precisely where you have slewed to, you have a deep map
of the field to compare the image with.

Another factor here is the availability of 0.33× telecompressors for f/10 Schmidt–Cassegrains from the US
company Optec. Not only does this give a three times
wider field of view, it also optimises Deep Sky imaging
so that the field is not being over-sampled (see Chapter
8 for more details).

My own 30 cm LX200 seems to slew to an accuracy
of around 5 arc-minutes when carefully polar aligned
and when a single star is used for calibrating the RA
and Dec. With the Optec telecompressor installed,
giving a 1 metre focal length, I enjoy a CCD field (SBIG
ST7) of 16′ × 24′. Slewing to an object is rarely hit and
miss with this arrangement.

I have to stress that if your LX200 is NOT well aligned
(either in Alt–azimuth or Polar modes), it will typically
mis-slew by the basic 5′ (or so) PLUS an amount up to
the size of your polar alignment error. So don't accuse
Meade of exaggerating their slewing accuracies – the
real cause is likely to be your polar alignment!

It is worth mentioning that the UK's leading supernova hunter, Mark Armstrong, uses a 25 cm Meade
LX200, remotely controlled from the comfort of an
indoor room. The ability of some modern SCTs to reliably slew to a position within arc-minutes of that
required is of paramount importance and the main
reason why Meade LX200s have become the telescope
of choice for many amateurs world-wide.

Summary

Overall, the *advantages* of Schmidt–Cassegrains are:

- Extremely compact and portable design from the
  biggest telescope manufacturers in the business.
- A remarkably convenient eyepiece position which
  varies little with object position.
- Many accessories available to simplify observing,
  astrophotography and CCD imaging.
- Ideal for people who like to play with gadgets as
  much as observing!
- Compatible with CCD autoguiders/cameras.

- High-precision slewing (unavailable with medium/ large aperture Newtonians) make SCTs the telescope of choice for Supernova patrolling.

The *disadvantages* are:

- More expensive than Newtonians, even for basic models.
- Very expensive in large apertures: 30, 35 and 40 cm models will cost 5 to 10 times as much as a Dobsonian of the same aperture.
- Instruments are manufactured in the USA – good news if you live in the USA, bad news if you live in the UK or Europe (where you could end up paying twice as much as Americans do).
- Lightweight, thin-mirror, fast-primary optics have not always been known for delivering the best planetary images. However, this criticism is less relevant than, say, ten or twenty years ago. Competition between the rival SCT manufacturers has noticeably improved SCT optics, but you will not achieve the definition that you would with an f/10 Newtonian of the same aperture!

Before the 1970s, it would have been impossible to predict that in the future SCTs would become vastly popular, but the compactness and portability of these instruments have been of fundamental importance, especially in an era where many observers want to haul their telescopes to less light-polluted sites.

Other, perhaps equally, important factors have been the remarkable growth in the power of electronics and home computers, and the commercial competition between Celestron and Meade. This has resulted in the simple and basic 20 cm SCT of the 1970s developing into a sophisticated electronic marvel of the 1990s. It is worth remembering that the autoguiding, precision slewing and CCD imaging capabilities available to the amateur SCT user would have been the envy of professional observatories only twenty years ago!

Equipment in amateur hands these days can image objects far fainter than the objects recorded in the original Palomar Sky survey: it is the Schmidt–Cassegrain that has been the test-bed for most of these innovations.

Maksutovs

The Maksutov is an alternative form of catadioptric telescope which combines the portability and compactness of the SCT, but is regarded by many as having superior planetary definition.

The general design of the Maksutov–Cassegrain telescope (to give its full name) is shown in Figure 3.8. At first glance the design looks very similar to the Schmidt–Cassegrain – but the subtle differences are all-important.

The primary mirror of a Maksutov consists of a strongly aspheric (typically f/2.5) primary mirror and a spherical secondary mirror which is actually an aluminised spot deposited on a spherical meniscus corrector lens. When you look at the corrector plate of an SCT, it looks more or less flat. The meniscus corrector lens of a Maksutov is very obviously *not* flat, it is steeply curved with its convex side towards the primary mirror. The aluminised spot in the middle of the corrector (on the inside, of course) typically multiplies the effective focal length of the primary mirror by a factor of six, resulting in (for example) an f/15 system at the focus. As with commercial Schmidt–Cassegrains, computer-optimised primary and secondary mirror baffles and primary baffle field stops are used to optimise the performance.

The Maksutov shares most of the advantages of the Schmidt–Cassegrain. The design is highly compact, rugged and free from tube currents. The optical characteristics are excellent. The design is free from coma and astigmatism and there are no diffraction effects from the secondary mirror supports (the "spider" on a simple Newtonian) as there are none. In addition,

Figure 3.8. The Maksutov.

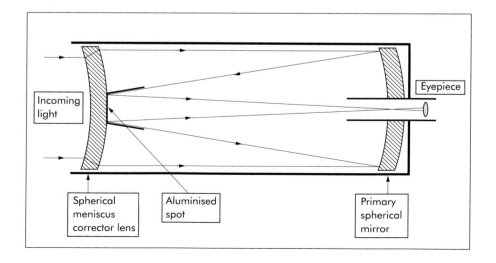

Incoming light

Eyepiece

Spherical meniscus corrector lens

Aluminised spot

Primary spherical mirror

collimation is not an issue because the optical elements of commercial Maksutovs are fixed and so collimation is guaranteed.

Unfortunately, because the corrector has to be ground from a very thick and expensive piece of optical glass, the design is expensive to manufacture in all but modest apertures.

For many years the US company Questar was the sole manufacturer of quality Maksutov telescopes. Their 90 mm models were remarkably portable instruments, highly prized by all who owned them. Their 180 mm model was even better, although prohibitively expensive. I can't even bring myself to mention the price of their 300 mm Maksutov!

In the 1980s a couple of ex-Questar employees set up a company called Quantum and set about producing Questar-like Maksutovs in 100 mm and 150 mm apertures. The company survived for a number of years and produced high-quality instruments. The only significant design difference when compared to Questar's Maksutovs was the use of a single equatorial fork arm on the Quantum models.

The prohibitive cost of Maksutovs meant that Quantum's life span was limited, for the benefits of Maksutovs over Schmidt–Cassegrains were not significant enough to sustain two US companies in this sector of the market.

Then, in the 1990s it was – once again – Meade who brought mass-produced Maksutovs back to the commercial market. Meade currently produce two Maksutov telescopes: the 90 mm Meade ETX and the 180 mm LX200 Maksutov. The $500 ETX has been a runaway success for Meade. The telescope features virtually all of the advantages of Questar's legendary 90 mm model, but at a fraction of the cost (see Figure 3.9, *overleaf*).

On its release in 1996, the Meade ETX sold so well that a "premium" market in ETX sales started to develop. Meade have now increased production to match demand. The 180 mm LX200 Maksutov is considerably more expensive but is still remarkable value for money, incorporating, as it does, all the fast slewing and precision pointing features of Meade's LX200 Schmidt–Cassegrains.

In this section I have introduced the reader to the most popular forms of amateur telescopes. There are other variants but they are rarely offered commercially and are largely the province of the ATM (Amateur Telescope Making) community.

Figure 3.9. The Meade ETX.

Schiefspieglers

In the UK there are fewer amateur telescope makers per head of the astronomical population than in the USA, but a few do exist.

Undoubtedly the best known "unusual" home-made telescope in the UK is the *Schiefspiegler* of Terry Platt, the UK's leading planetary imager and the "brains" behind the Starlight Xpress range of CCD cameras (Figure 3.10, *opposite*). Prior to Terry's work the UK only had one renowned planetary photographer, the late Horace Dall (1902–86), who I have already mentioned. Horace died just before the CCD era, whereas Terry's reputation has risen in parallel with it. Indeed, in the UK, Terry started the revolution, for which he

Figure 3.10. Terry Platt, the brains behind Starlight Xpress CCDs.

was appropriately awarded the British Astronomical Association's "Horace Dall Medal" in 1997.

The Schiefspiegler is a reflecting telescope which has no obstruction in the light path; thus, it combines the advantages of a refractor (unobstructed aperture) with those of a reflector (larger affordable apertures with no chromatic aberration). In fact, because the light path is folded, the telescope can be made a convenient size, despite its long focal length. Figure 3.11 (*overleaf*) shows the design of Terry Platt's Schiefspiegler.

The original Schiefspiegler design was pioneered by Anton Kutter and Dick Buchroeder in the 1950s. It consisted of a long focal length concave primary mirror with a convex, tilted, secondary mirror positioned off to one side of the primary's optical axis; thus, not obstructing the incoming light. The primary and secondary mirrors should have the same radius of curvature; but the former is concave and the latter convex. This original design only worked for relatively small apertures, but Buchroeder added a third, weakly concave, "tertiary" mirror (in the 1980s) enabling the design to be used for apertures up to 300 mm or so. This variant is known as the *Tri-Schiefspiegler*, but Terry Platt's instrument is, strictly speaking, a "*Quad-Schiefspiegler*" as it uses a fourth flat mirror to create a more convenient eyepiece position and a final image which is not laterally inverted (i.e. not a mirror-image).

Terry's instrument, at 318 mm aperture, is as large as a Tri-Schiefspiegler can be. The primary mirror is

f/12 and the final f/ratio is f/20. Using CCD cameras (of his own design) with 15 micron pixels and a barlow lens (to give f/40), Terry has secured stunning planetary images from the UK, a country renowned for its chronic atmospheric seeing. It should be stressed that building a telescope of this complexity demands skill, determination and, above all, patience.

Figure 3.11. Terry Platt's Quad-Schiefspiegler, from the primary end.

Few telescope makers have attempted to make Schiefspieglers, and fewer still the CCD camera to go with it. But, if you think the Schiefspiegler is what you need, but will never be able to make one, don't imagine that high-quality CCD imaging is out of your grasp. Although the Schiefspiegler has an unobstructed aperture and therefore superb planetary definition, there is no evidence that it can out-perform a long-focus Newtonian of the same aperture. A 320 mm f/7 Newtonian with a good mirror and a small secondary should perform just as well as a 320 mm f/20 Schiefspiegler.

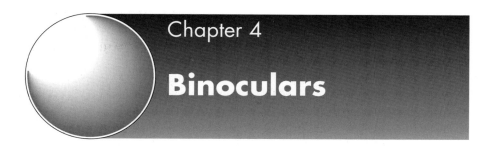

Chapter 4

Binoculars

To most beginners in Astronomy, binoculars seem at first to be of very limited use. Surely a telescope is what is required to enjoy astronomy or do useful scientific work? Nothing could be further from the truth.

In terms of visual enjoyment only the most versatile of small alt–azimuth telescopes can match a decent pair of hand-held binoculars. Even small-aperture binoculars will greatly enhance your view of the night sky in comparison with the naked-eye view, and sweeping the Milky Way with large binoculars can be breathtaking. With any form of telescope there is, inevitably, a period of set-up time and some hassle in identifying the small and inverted field of view. And the eyepiece of a Newtonian always seems to be in the most inconvenient position! No such problems exist with binoculars; you simply pick them up, stroll outside and aim them by hand.

Obviously there is a limit to the weight and magnification which you can comfortably support, but beyond this point you can buy a versatile binocular-stand to relieve your aching arms.

I said in Chapter 1 that the pupil of the eye can only dilate to 7 or 8 mm in young people, and by middle age 6 mm is a more likely limit. Most observers will find that – with large binoculars – a magnification of about 15× is the most that can be comfortably hand-held for periods of even a few minutes. The result is that a practical upper limit for hand-held binoculars is about 15× magnification and 90 mm aperture, or "15 × 90" in binocular terminology. Such a pair of binoculars will give impressive wide-field views but unless you are built like Arnold Schwarzenegger, you will succumb to aching arms and shoulders!

Most binocular eyepieces have apparent fields of view of around 50° so 15× magnification will give a real field on the sky of about 3.3°. Although 15× magnification is a practical upper limit, most people prefer a lower magnification of say, 10× or so. In fact, 10 × 50 binoculars are an excellent choice for amateur astronomy.

Ultra-small, highly portable binoculars are becoming highly popular in high street shops but they are not very useful for astronomy and are primarily designed for compactness and portability. Typically, they have apertures of around 25 mm.

The legendary British comet and nova discoverer (five of each), George Alcock, has made every one of his ten discoveries with binoculars. Most of his comet discoveries were made with tripod-mounted ex-military 25 × 105 binoculars (see Figures 4.1 and 4.2, *opposite*). His first four Nova discoveries were made with 15 × 80 hand-held *Beck–Tordalk* binoculars; the fifth, in 1991, with 10 × 50 binoculars. The Beck–Tordalk 15 × 80s were also employed for the discovery of Comet Iras

Figure 4.1. George Alcock – living legend!

Figure 4.2.
Alcock's 25 × 105 ex-World War 2 binoculars.

Araki Alcock in May 1983. Although now in his mid 80s (he was born on August 28th 1912) George continues to sweep the night sky but now prefers the 3 mm exit pupil of his Russian 20 × 60 binoculars. The higher magnification and requirement for a rock-steady pair of hands seems to cause George few problems!

Celestron, Fujinon and Swift are among the leading manufacturers of astronomical binoculars and the potential buyer is spoilt for choice. I have already recommended 10 × 50 binoculars for astronomy; the 5 mm exit pupil will suit young and middle-aged observers alike and 10 × 50 binoculars are widely available and (relatively) inexpensive. They are also a lot lighter than 70 mm or 80 mm models. The Russian-made *Helios* range of binoculars is certainly worth looking out for; although not as attractive and compact as the top manufacturers' models, they are remarkably good value for money and the optics are of good quality.

Of all the binoculars available to the keen young observer I would rate Celestron's Ultima 9 × 63 model among the very best. Although the 7 mm exit pupil may be a challenge for even some young eyes, the magnification is ideal for hand-held work and the light weight of these binoculars is a definite bonus.

If your pupils only dilate to 5 mm, a pair of 9 × 63s will perform as if they were 9 × 45s, but allow your pupil to roam a bit within the generous 7 mm exit pupil!

Stands for Binoculars

After a few minutes observing with heavier binoculars, the amateur astronomer will search for a convenient surface to rest his or her elbows on. A low wall can be very useful here, as can a sun-bed with arm rests. I found the bonnet and roof of an Opel Corsa car to be ideally positioned for reclining on/resting my elbows on while observing Comet Hyakutake!

Large binoculars usually come supplied with a metal strut which can fix the binocular body (at the hinge screw) to a standard photographic tripod head. Larger binoculars may well come ready-supplied with an observing pillar or tripod. Perhaps the ultimate example of commercial astronomy binoculars are Fujinon's 25 × 150s (Figure 4.3). These massive binoculars come with a $10,000 price tag but are particular favourites among Japan's most successful comet

Figure 4.3.
Fujinon's 25 × 150 binoculars.

searchers. Comet 1996 B2 Hyakutake was swept up using a pair of Fujinon 25×150s.

A useful feature of these and some other large binoculars are the 45° angled eyepieces, enabling an observer to adopt a convenient viewing position when sweeping the night sky. An array of ingenious binocular mountings are available for those observers who want to relax in comfort while scanning the skies. A list of binocular mounting suppliers is given in Appendix 1.

Most binocular mount designs consist of a counterbalanced parallelogram support, not dissimilar in appearance to the mounting of an "anglepoise" lamp. Stability is an all-important feature and attaching a binocular mount to a flimsy tripod is a recipe for a shaky image and a smashed pair of binoculars (when the mounting topples over).

One system – called the *Sky Hook* – allows the user to recline on a sun-bed while using binoculars. Now that's what I *call* observing! You can even purchase binocular "chest mounts" which use a support structure on the observer's chest to support the weight of the binoculars. The models I have seen can only be classed as partially successful and have the added disadvantage of making the wearer look like an idiot!

In my opinion the most successful and comfortable way of observing with powerful binoculars and avoiding arm shake is to invest in binoculars with a stand and 45° angled eyepieces. Unfortunately, you have to buy fairly large-aperture binoculars to achieve this degree of comfort.

In the UK a pair of Miyauchi 20×77 binoculars with a 45° viewing angle can be obtained for an affordable price from True Technology (Figure 4.4, *overleaf*). A good tripod mount can then be added for considerable observer comfort up to altitudes of 45° or more. Extremely compact Russian-made 15×110 binoculars can also now be obtained (Figure 4.5, *overleaf*).

Image-Stabilised Binoculars

A recent innovation in binocular design, of particular interest to amateurs, are *image-stabilised* binoculars.

Image-stabilising techniques have been used for some years in the camcorder industry, and in the

Figure 4.4.
Miyauchi's 20×77
binoculars.

sophisticated field of active/adaptive optics. Basically, a microprocessor system makes a judgement as to whether rapid image shake of a scene is taking place, and if it is then tiny actuators/motors move lightweight optical components to compensate.

Needless to say, a lot of this technology was originally developed for military applications and as is so often the case, eventually filtered down to the consumer market. Even so, I find it remarkable that

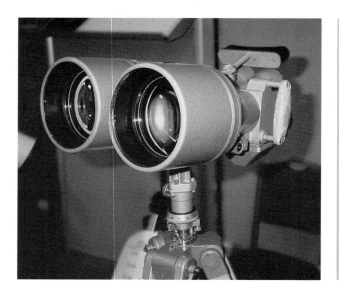

Figure 4.5.   Russian
15×110 binoculars.

"affordable" image-stabilising binoculars are already with us. (Note that an example of "affordable" in this context is $1100 for Canon's 15×45 models.) At the limit of hand-held observation, image stabilisation can make small binoculars perform like bigger models.

Although Fujinon's 25×150 binoculars are, perhaps, the largest that most amateurs would wish to own, Amateur Telescope Makers (ATMs) who crave viewing the night sky in "stereo" have built enormous "double telescope" binoculars of 30, 40 and even 50 cm aperture. Building such monsters is beyond the scope of this book but articles on their construction can regularly be spotted in *Sky & Telescope* magazine.

Chapter 5

Eyepieces

Magnification

Whenever I show non-astronomers pictures of my telescopes, the immediate reaction is predictable, something like, "Wow, what kind of magnification can you get with that?". Photographs of the Moon produce the same response. If I'm in a picky mood, my unhelpful response might be, "Well ... that crater is about sixty miles across and it's only an inch across on the print, so it's a *reduction* and not a magnification".

By *magnification* an astronomer is talking about the *angular* increase in size seen by the eye. Figure 5.1. shows the relationship between eyepiece, exit pupil, focal plane and field stop. The Moon appears about 0.5° in diameter to the naked eye, and if it appears 30° wide through the eyepiece then the magnification is

Figure 5.1.
Telescope, eyepiece, field stop, focal plane and exit pupil.

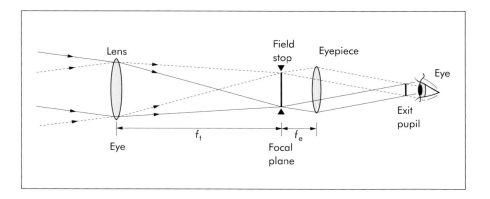

60×. Fortunately, calculating magnification is simple: just divide the focal length of the telescope by the focal length of the eyepiece.

The *apparent field* of an eyepiece is fixed by the eyepiece design and construction. The widest apparent fields available with today's eyepieces are over 80°, enough to put the edges of the field at the edges of the eye's field of view and make the observer feel as if he is floating in space!

At 160× with an 80° apparent field, the real field on the sky is 0.5° – that is, you can study tiny craters on the Moon, just a mile or two in diameter while still getting the whole Moon in the field – awesome!

So what range of magnifications is desirable to get the best out of a telescope?

Lowest Practical Magnification

I have already explained (in Chapter 1) why an exit pupil greater than 7 mm in diameter (less than 10× with a 70 mm telescope) marks the *lowest* practical limit. Any lower magnification, and all the light cannot enter even the youngest dark adapted pupil. A handy formula to remember here is:

(required eyepiece focal length) = (telescope f/ratio) × (desired exit pupil diameter)

Thus the longest (7 mm exit pupil) focal length eyepiece required for an f/5 telescope is: $5 \times 7 = 35$ mm. For an f/7 telescope and a 7 mm exit pupil, an eyepiece of 49 mm focal length would be required.

Highest Practical Magnification

At the other end of the scale, I have said that atmospheric seeing is the limiting factor. But, to be prepared for the best *seeing* (atmospheric transparency and stability), it is good to know what is the highest magnification that could ever be required.

The naked eye of a person with perfect vision has a resolution of about 1 minute of arc (1/60th of a degree, or 60 seconds of arc). Less fortunate observers can resolve around 2 minutes of arc.

A 30 cm telescope can theoretically resolve 0.4″, according to the Dawes criteria; in practice, detail smaller than this is virtually never seen, even with very large instruments, which is why dedicated amateurs can compete with professionals where planetary imaging is concerned. A simple calculation shows that if the resolution of the eye is taken as 2 minutes (or 120 arc-seconds) and the resolution limit of a 30 cm telescope is 0.4″, then *all* the detail should be visible with a magnification of 120/0.4 = 300. This is equivalent to 10× per centimetre of aperture, which is a convenient rule-of-thumb to remember.

Older textbooks often quote "50 × per inch of aperture" as the maximum magnification, which is 20 × per centimetre, or double my "rule-of-thumb" above. Well, yes – in perfect seeing, huge magnifications can probably be justified; they ensure that the observer is not straining to see the detail at the limit. However, I prefer the 10× per cm rule, which at least ensures that lunar and planetary images are reasonably bright. A lot depends on the preference of the individual observer and his eyesight, so fixed rules are not really appropriate here. Most observers will soon settle for their own "favourite" high-power eyepiece.

Now, 10× per cm translates to an exit pupil diameter of 1 mm so taking the 1 mm exit pupil as a baseline, we can go back to my earlier formula to work out the highest power eyepiece that a "typical observer" will need:

$$\text{(required eyepiece focal length)} = \text{(telescope f/ratio)} \times \text{(desired exit pupil diameter)}$$

Thus to provide a 1 mm exit pupil with an f/5 system, a 5 mm eyepiece is required. For an f/7 system, a 7 mm eyepiece is required: for an f/10 system (typical for a Schmidt–Cassegrain) a 10 mm eyepiece would be needed.

Practical Considerations

In practice, most observers will need a minimum of three eyepieces – for low-, medium- and high-magnification observing.

If you have a tight budget and want to get by with only three eyepieces, my recommendation is to settle for eyepieces that give exit pupil sizes of 6 mm, 2 mm and 1 mm with a telescope of respectable aperture (say, 20 cm up). For a 30 cm telescope this translates into magnifications of 50×, 150× and 300× respectively.

A magnification of 50× will give a 1° field with a modern eyepiece, which is ideal for observing large Deep Sky objects and the whole Moon. And a 6 mm exit pupil is less of a challenge to the dark-adapted middle-aged eye than 8 mm is!

The resultant – slightly higher – magnification makes the background sky noticeably darker too. Magnifications of 150× and 300× will be ideal for observing the Moon and planets under typical and excellent conditions respectively.

When buying eyepieces it is also worth checking if the focal lengths you require are available as a *parfocal set*. "Parfocal" means that the telescope does not need re-focusing for each eyepiece (which can be a real hassle if you are constantly switching eyepieces in the dark).

Comet Seeking

This is a fitting point at which to mention the concept of the "Comet Seeker" telescope. For over two hundred years amateur astronomers have competitively hunted for comets with special wide-field visual telescopes. A survey of the instruments and magnifications used by these comet hunters reveals their liking for an exit pupil of around 6 mm, or around 1.66× per cm of aperture. The remarkable Australian comet hunter William Bradfield, discoverer of no less than *seventeen* comets, frequently uses a 15 cm f/5.5 refractor at a magnification of 25×. David Levy used a much larger telescope – a 40 cm Dobsonian – for his eight visual discoveries but, at 60×, he too prefers a 6 mm exit pupil.

The reason for such a choice is not hard to see. Comet hunters want as much light as possible to enter the eye so that faint comets are easily detected. But they also want a wide field so they can cover plenty of sky as quickly as possible. A 6 mm exit pupil provides the lowest practical magnification for the middle-aged observer, and more contrast than an 8 mm exit pupil.

The jury is still out on whether small or large telescopes are better for comet sweeping. Small telescopes of 15 cm aperture cover the sky more quickly, but discoveries fainter than mag. 10 are rare. Conversely, 40 cm reflectors enable discoveries to mag. 12 but take an age to cover the sky.

Focusing-Tube Diameter

For the widest fields there is another decision to make: whether to use a standard $1\frac{1}{4}$-inch (31.7 mm) eyepiece or a 2-inch (50.8 mm) eyepiece (Figure 5.2).

The advantage of a 2-inch eyepiece is that the internal barrel diameter is approximately 46 mm, compared with just 27 mm in a $1\frac{1}{4}$-inch diameter eyepiece. Eyepieces with long focal lengths (i.e. low magnifications) and wide apparent fields will have their fields compromised by the smaller barrel diameter.

Why? Let's begin by making the approximation that the eyepiece barrel merely needs to accommodate the diameter of the focal plane image. From this a useful formula can be derived which will tell us what minimum size of eyepiece barrel is needed. If D is the diameter of the image formed at the telescope focal plane (in millimetres) then:

$$D = F_A \times F_L/57.3$$

where F_A is the apparent field of the eyepiece in degrees and F_L is the focal length of the eyepiece in millimetres.

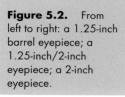

Figure 5.2. From left to right: a 1.25-inch barrel eyepiece; a 1.25-inch/2-inch eyepiece; a 2-inch eyepiece.

Rearranging this formula and making D equal to 27 (the maximum internal diameter of a $1\frac{1}{4}$-inch eyepiece) gives us the maximum focal length eyepiece (MFL) that we can use with the smaller eyepiece diameter:

$$\text{MFL} = 27 \times 57.3/F_A$$

Thus for 31.7 mm eyepieces a 50° apparent field design implies a MFL of 31 mm; a 65° apparent field design implies a MFL of 24 mm and an 84° apparent field design a MFL of 18 mm.

As the eyepiece apparent field and focal length alone determine the barrel size requirement, it is not necessary to take the design of the telescope into consideration.

A brief glance through the catalogues to see the available eyepieces on the market confirms this. For example, Tele Vue's 68° field *Panoptic* range eyepieces switch to a 2-inch barrel for focal lengths above 27 mm, and their 82° Nagler range eyepieces switch above 16 mm. In practice, the maximum real fields available with today's eyepieces (filling a 2-inch barrel field lens) are found in the 55 mm 50° field Plossls, or 40 mm 67° Super Wide Angle eyepieces.

But before you rush out and buy a 55 mm 50° eyepiece, remember what I said earlier about exit pupils. Unless your telescope has an f/ratio of 7 or more the exit pupil will exceed 8 mm – the limit for even the youngest eye!

Real Field Limitations

To calculate the maximum possible real field ($x°$) on the sky for a given telescope focal length, the following formulae are sufficiently accurate for the small angles involved.

For 1.25-inch eyepieces:

$$x° = (27 \times 57.3) / (\text{focal length in mm}) = 1547 / (\text{telescope focal length in mm})$$

For 2-inch eyepieces:

$$x° = (46 \times 57.3) / (\text{focal length in mm}) = 2636 / (\text{telescope focal length in mm})$$

Now, having had a look at some eyepiece maths, let's see what is actually available in practice.

Popular Commercial Eyepieces

In the last fifteen years, the performance of eyepieces has improved dramatically, owing to advanced ray-tracing software, new types of glass and intense market competition. In the 1970s the amateur had to choose between "cheap and nasty" eyepieces (Ramsdens, Huygenians and Kellners) or Orthoscopics.

The wealthy 1970's amateur might possibly consider an Erfle for wide-field observing; the dedicated planetary observer might dream of acquiring a Tolles or Monocentric eyepiece for planetary observing.

The cheap eyepieces had fields of 30° or 40°; the Orthoscopics 40°; the Erfles, 65°. In addition, the cheapest eyepieces only worked well at f/ratios slower than f/7 or f/8 (hence their popularity with manufacturers of 60 mm f/12 refractors!) and suffered from poor colour correction (very poor in the Ramsden design). OK, to be honest, the choice twenty years ago was a bit wider than that, but not much.

In the 1980s Al Nagler of Tele Vue introduced his high-quality Plossl and ultra-wide-field Nagler designs to the amateur marketplace and things have never been the same since. The Plossl is not a new design but was generally considered less attractive than the Orthoscopic eyepiece, prior to the 1980s. However, today's Plossls can boast wider fields (50°) than the traditional orthoscopic eyepiece as well as equivalent sharpness in the centre of the field (Figure 5.3).

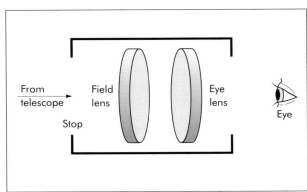

Figure 5.3. The Plossl eyepiece.

Once considered the "Rolls Royce" of eyepieces, Orthoscopics are still available (e.g. from Orion Telescope Center) but the wider field of the modern Plossl makes it the favourite choice for today's amateurs. Plossls and Orthoscopics work well down to f/5 or even f/4 and are available from a variety of sources such as Tele Vue, Meade and Celestron. Celestron also market the impressive Ultima range of eyepieces in this category; these are of a hybrid four- to seven-element design and feature multi-coated optics, long eye relief and minimal aberrations.

Once you become used to today's wide eyepiece fields of 50° and above, changing back to an older style of eyepiece makes you wonder how you tolerated the lesser models. Recently, I removed a modern eyepiece and looked through one of my earlier models from the 1970s with the distinct feeling that I was peering down a ball-pen refill, *not* an eyepiece!

Moving up to wider fields: above 50° we next come to eyepieces with fields of around 65°, i.e. the region once dominated by the Erfle. Whereas (1998 prices) the Plossls occupy the $40 to $90 price range (depending on whether you want just a very good or an exceptional eyepiece) the 65° eyepieces command prices of up to $250 for $1\frac{1}{4}$-inch barrels and up to $350 for 2-inch barrel designs. So we are talking about a five-fold price increase for a 70% increase in field area! This category is dominated by Tele Vue and Meade.

Meade's Super Wide Angle eyepiece range boasts a 67° field with focal lengths from 13.8 mm to 24.5 mm for $1\frac{1}{4}$-inch barrels and 32 or 40 mm for the 2-inch barrel. Tele Vue's equivalent Panoptic (68° field) eyepiece range features 15 and 19 mm eyepieces in the small barrel size, a 22 mm dual-barrel eyepiece and 27 or 35 mm large barrel eyepieces. If you want a wide field with stars sharp to the edge, even at f/5, these eyepieces are for you; but you may need to consult the bank manager first!

Not surprisingly, the same two companies slog it out in the ultra-wide field eyepiece category too; with 82°–84° fields the observer has to press his eye right up to the eye lens and even turn from side-to-side to see where the field ends!

Tele Vue's Nagler's were the first mass-produced eyepieces to offer such extraordinary wide fields. Needless to say, at up to $400 per eyepiece, few amateurs possess the full set! Without a doubt the most remarkable eyepiece in this category (or any category perhaps!)

is Tele Vue's 20 mm Nagler (Figure 5.4). Standing 130 mm high and nearly 70 mm in diameter (2-inch drawtube fitting) this 1 kg eyepiece is a monster which will have you re-balancing your telescope to cope with the extra weight! With a short-focus 25 cm f/4 Newtonian, this eyepiece will deliver 50× magnification with a 1.6° field and a 5 mm exit pupil.

Incidentally, to verify the enormous field of view of this (or any) eyepiece is relatively simple. Point your telescope at a star at about 0° Declination and then turn the drive off; then time how long it takes for the star to drift from one edge of the field to the other, in seconds (t). The apparent field of view of the eyepiece is (15× magnification × t) / 3600. (If you choose a star above Declination +10° or below Declination –10° you will need to multiply the result by the cosine of the Declination to get an accurate figure.)

Naglers, like any eyepiece, have their field of view limited by a field stop (a black ring inserted into the eyepiece barrel at the position of the focal plane). Beyond this point unacceptable optical aberrations exist. The observer sees a sharp black edge to the field of view when looking through the eyepiece owing to the field stop. For many designs this field stop is located in the chrome drawtube barrel of the eyepiece, but in the

Figure 5.4.
TeleVue's massive 20 mm Nagler dwarfs a standard eyepiece!

complex Nagler design it is inside the seven-element optical lens assembly. Tele Vue's Nagler range extends from 4.8 to 20 mm; Meade's competing Ultra-Wide angle range covers 4.7 to 14 mm. With a 50 mm barrel, 30 mm focal length eyepieces with 80° apparent fields, are technically possible but would be prohibitively expensive. With 67° apparent fields available up to 40 mm focal length, the market for ultra-wide field eyepieces of more than 14 mm focal length must be small.

Both Tele Vue and Meade's ultra-wide angle eyepieces are miracles of modern optical technology. The hardest eyepieces to design and build are those that will work at fast f-ratios, cover wide fields and remain ultra-sharp at the centre; these eyepieces deliver all three requirements – remarkable!

Tele Vue, Meade and Celestron are not the only manufacturers of quality eyepieces. Vixen's LV Lanthanums, Clavé's Plossls and the Pentax SMC-XL range are all excellent eyepieces, and scanning the equipment reviews of *Sky & Telescope* and *Astronomy* magazines will help you to decide which eyepieces are best for you. Vixen's LV Lanthanum range (Lanthanum is a rare-earth glass permitting special optical designs) covers 2.5–30 mm focal lengths, all with a 20 mm eye relief (exit pupil to eye lens distance); a useful feature, especially if you have to wear spectacles while observing.

In 1997 Tele Vue introduced a novel *zoom* eyepiece with variable focal length from 8 to 24 mm; if you get tired of fumbling around for eyepieces in the cold and damp (and occasionally dropping them!) this could be the eyepiece for you.

High-Definition Eyepieces

Some observers have little interest in wide fields of view, preferring instead the ultimate in high-definition for lunar and planetary observing. For such observers the rarely seen Monocentric or Tolles eyepieces are the ultimate possession (in many of their opinions).

For high magnifications with moderate focal lengths, very short focal length eyepieces are required.

Manufacturing the small glass lenses which are required to reveal faint and subtle planetary markings at fast f-ratios is highly difficult and with today's eyepiece emphasis on wide fields some planetary observers prefer to acquire specialist narrow-field eyepieces. It should be pointed out here that, for planetary observing, eyepieces will always perform better at longer f/ratios and focal lengths. My own 36 cm Cassegrain/Newtonian delivers far superior planetary images at f/25 than at f/5 and also out-performs my 49 cm f/4.5 Newtonian. If a Newtonian is being purchased or built for planetary observing, an f/ratio of 6 or slower and a focal length approaching 2 metres is highly desirable. For those observers who still believe that the key to seeing fine planetary detail is the right eyepiece, the two specialist eyepieces for this type of work are described below.

The Tolles eyepiece (Figure 5.5) is, essentially, a cylinder of solid crown glass with curved ends: a solid version of the Huygenian eyepiece in fact, with a similar 30° field. Owing to the solid nature of the design the field of view is free from ghost images and internal reflections. In fact many reflections attributed to poor eyepiece design are often aberrations within the observer's own eye lens.

The eye relief of the Tolles is very poor (disastrous for spectacle users) but the transmission of light is 85% and central definition is excellent. The Tolles is best used with telescopes of f/7 or slower.

The Monocentric eyepiece (Figure 5.6, *overleaf*) is similar to the Tolles in that it has no internal air spaces, although it is not actually a solid lump of glass. It is, in fact, a cemented triplet and the lens surfaces are parts of concentric spheres. The Monocentric eyepiece has far better eye relief than the Tolles but a slightly smaller field of view, typically 25°. Light transmission can be as high as 90% with coated optics and the design is usable down to f/6.

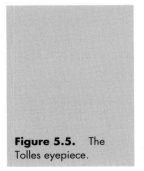

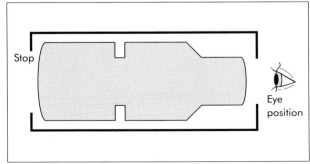

Figure 5.5. The Tolles eyepiece.

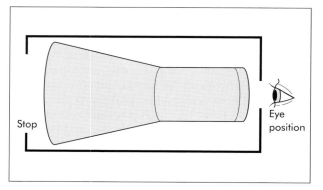

Stop

Eye position

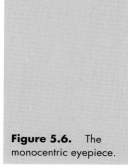

Figure 5.6. The monocentric eyepiece.

Barlow Lenses

You may have gathered from this discussion of eyepieces that they all perform better at the slower f-ratios. Unfortunately, telescopes' f-ratios are, essentially, fixed and so the users of short-focus Newtonians may despair of ever achieving the best planetary views. Fortunately, help is at hand in the form of the *Barlow lens* (Figure 5.7).

The Barlow lens is a simple looking device that usually consists of a diverging achromatic doublet or triplet lens mounted at the end of a chrome tube and a longer metallic extension to this tube into which eyepieces can be inserted. The action of the diverging lens is to increase the effective focal length of the telescope by an amount which is determined by the diverging power of the lens and the separation between it and the

Figure 5.7. The Barlow lens.

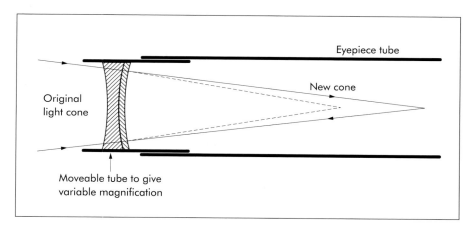

Eyepiece tube

New cone

Original light cone

Moveable tube to give variable magnification

eyepiece. The greater the separation, the greater the *amplification factor.*

The amplification factor can be determined by dividing the lens-to-new-focus distance by the lens-to-old-focus distance. Barlow lenses are designed to operate at one specific amplification factor, usually 2×. Currently there are commercial models offering 1.8×, 2× or 2.5×, with fixed tube lengths to guarantee these amplification factors. However, for each lens design, optical aberrations can only be minimised for a single amplification factor. There is little point using a Barlow lens on a fast Newtonian to achieve better eyepiece performance if the Barlow lens itself introduces more aberrations!

For users of fast reflectors with 2-inch drawtubes one can even obtain 2-inch barrel "big" Barlows. A possible application for such a device might be the owner of an f/4 Newtonian who wished to have a low-power field but with sharp star images. A "big" 2× Barlow would deliver f/8 quality and enable large-barrel 55 mm or 40 mm focal length eyepieces to be used, yielding 7 mm and 5 mm exit pupils respectively.

Eyepiece Projection

Another useful application of Barlow lenses is as an alternative to *eyepiece projection* work. We have already seen that for photography or CCD imaging of the Moon or planets a large focal length is required; typically 10 metres with a CCD camera and – maybe – 30 metres with film. To achieve such long focal lengths, eyepiece projection is often used. This involves using an eyepiece to project a magnified image onto the CCD or film. The eyepiece is placed outside the normal visual focus position such that the emergent rays are not focused at infinity (for the relaxed eye) but focused at the film/chip focal plane.

As with visual observing, eyepieces used for projection purposes with fast telescopes are subject to aberrations which may well prevent the finest details from being recorded. In addition, when used for projection, eyepieces are operating outside their optimum design range; they are, after all, designed for visual use.

Some dedicated planetary workers use high-quality microscope objectives for projecting the planetary image, but a simpler alternative is to use a Barlow lens. A standard Barlow will double the effective focal ratio of the telescope, reducing the aberrations when eyepiece projection is used. Most major manufacturers supply eyepiece projection units which feature a chrome-plated 1.25-inch (31.7-mm) external barrel for insertion into the drawtube or Barlow tube and a 42 mm T-thread for attachment to the CCD camera or 35 mm-format camera body. The amount of magnification generated by eyepiece projection can be calculated from the (simple!) formula I/O where O is the distance of the eyepiece from the telescope's focal plane and I is the distance from the eyepiece to the final image. However, in practice, determining these distances can be tricky and trial and error is the method employed by most amateurs.

Illuminated-Reticle Eyepieces

The subject of guiding during long-exposure astrophotography has already been touched on earlier. However, eyepieces used specifically for guiding were not covered. In today's world of autoguiding, periodic error correction, CCDs and "track-and-accumulate" imaging, guiding eyepieces are less essential than they were ten years ago. However, for wide-field astrophotography of bright comets like Hyakutake and Hale–Bopp, small CCDs and the narrow fields of Schmidt–Cassegrains are inappropriate. With really bright comets, a small telescope plus a guiding eyepiece can be used as a highly efficient guidescope for a piggy-backed telephoto lens, with the observer guiding on the comet's nucleus.

Anyone who has attempted to guide a long-exposure astrophoto is well aware that seeing a star in an eyepiece is one thing; centring it on a cross hair and guiding on it is another! The faintest stars visible in a 60 mm refractor are around mag. 10 or 11, but the astrophotographer soon learns that stars fainter than mag. 6 will be almost impossible to guide on with a small aperture. Guide stars have to be visible with *direct*, not averted vision. The brightness of the guiding reticle should be adjustable to very low levels of illumination.

The human eye is not (strangely enough) very good at staring directly at one point for long periods. Try staring constantly at the full stop at the end of this sentence and you will see this is true. The brain likes a bit of movement, but when it is forced to stare at a star and a cross hair for many minutes, strange things start to happen! The brain can easily make the star disappear if the view does not change, especially if the reticle illumination is too bright. Some observers therefore prefer a cross hair that flashes on and off while guiding.

For this reason, Rigel Systems designed a guiding light source called "Pulsguide", a simple but effective invention that pulses the reticle illumination of a guiding eyepiece during an exposure. On and off times and the reticle brightness are all variable and the unit is compatible with many existing guiding eyepieces. For guiding on very faint guide stars, I find this gadget invaluable.

For guiding on a faint comet, the technique is to guide on a bright star but to slide the star along an illuminated, graduated scale to allow for the comet's motion. Therefore the scale should be adequate at the resulting magnification to yield a smooth, not a jerky, star trail on the final photographic print. Obviously the guiding eyepiece needs to be adequate to guide at the focal length of the eyepiece in use.

Observers with keen eyesight will find that a guiding magnification equal to the focal length of the astronomical camera ("astrograph") in centimetres will be adequate. This means that a photograph at the prime focus of a 36 cm f/5 Newtonian will warrant a guiding magnification of 180× (Figure 5.8).

Figure 5.8. The author's 36 cm f/5 Newtonian uses a high-power 12 cm refractor for guiding.

At the other end of the scale, a 300 mm focal length lens can be guided with a mere 30× magnification. Observers with less than perfect eyesight may prefer higher powers.

Commercial Guiding Eyepieces

A range of commercial guiding eyepieces to suit most requirements is available from the major manufacturers. Focal lengths of 12 or 9 mm are the most common, and a variety of different reticles is available. These days most of the available illuminated-reticle eyepieces are available in models that run on self-contained, coin-sized silver–zinc or lithium cells.

Although battery life is typically several hours, it is all too easy to leave an illuminated reticle switched on after an observing session (in which case the batteries last for exactly one session!). Alternatively, modern Schmidt–Cassegrains (such as Meade's LX200) feature a reticle power socket on their control panels; the corresponding Meade guide eyepieces have a 2 metre cable and plug.

Meade's 9 mm *Series 4000 Illuminated Reticle* is advertised as "The World's Most Advanced Illuminated Reticle Eyepiece". The novel feature of this eyepiece is the pair of milled knobs on the eyepiece barrel which allow micrometric centring (left–right, up–down) of a guide star without the necessity of repositioning the guidescope in its rings or slewing the telescope. Meade also offer a less expensive 12 mm eyepiece without the micrometric adjustment feature.

Celestron's *Microguide* eyepiece is more than just an illuminated reticle – a whole host of useful scales are visible in the eyepiece field. The 12.5 mm focal length *Microguide* has four separate scales at the focal plane:

1. An outer scale, useful for determining position angles on the sky.
2. A semi-circular position angle (PA) scale for measuring position angles of, for example, double stars.
3. A bull's-eye guiding reticle.
4. A linear scale for offset comet guiding or measuring the separation of the components of a double star. This scale is graduated at 100 micron (0.1 mm) intervals (equivalent to 10.4″ at a 2 metre focal length (160×).

The Celestron *Microguide*, designed at the Baader Planetarium in Germany, is a versatile eyepiece for a variety of applications, although some may find the wealth of glowing red data in the eyepiece field somewhat distracting and a limit to the faintest guide stars which can be used. Also, I personally found the outer PA scale somewhat blurred by aberrations at the field edge. Nonetheless, for offset guiding on comets and micrometric measurements it is a very useful gadget.

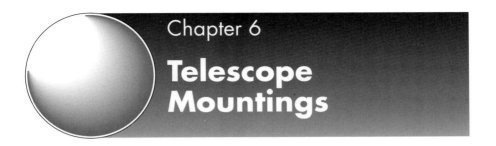

Chapter 6

Telescope Mountings

Most amateur astronomers buy their telescopes as a complete package, that is, the optics, tube, equatorial mounting and tripod are purchased as a working system.

However, specialists (e.g. planetary observers) often want something more, and custom systems comprising hybrid combinations of mountings and tube assemblies may often suit them better. In this chapter I'll talk about the standard equatorial mounting and drive systems supplied by the major manufacturers, as well as the "independent" commercial mounting systems and mountings which amateur telescope makers (ATMs) can build themselves.

The Equatorial Mounting

Astronomical telescopes are almost invariably mounted so that they have two distinct axes of movement, allowing the telescope to be aimed at any point in the sky. An equatorial mounting is simply a mounting in which one axis is parallel to the Earth's axis of rotation. This axis (which by definition must point to the north or south celestial pole) is known as the *polar* axis. The other axis, perpendicular to it, is called the *declination* axis.

The stars in the night sky appear to rise in the East and set in the West because the earth rotates with respect to them once every day (actually every 23h 56m 4s, but let's assume it is a day at this stage!). If the polar axis (remember it is North–South and therefore parallel to the Earth's axis of rotation) is rotated once per day in the opposite direction from that in which the Earth itself is turning, then the telescope should track the stars so that astronomical objects remain fixed in the field of view – just as if the Earth were not rotating at all.

In the northern hemisphere the stars rotate in an anti-clockwise direction around the north pole; in the southern hemisphere they rotate in a clockwise direction around the south pole.

During the last two hundred years or so – in fact ever since telescope drives became possible – various solutions to the problem of designing an ideal equatorial mounting have been devised.

I think it is fair to say that when a novice first encounters an equatorial mounting in the field, confusion still reigns! It certainly confused me as an eleven year old when I first encountered the school's equatorially mounted 76 mm refractor! I discovered that pointing the telescope at an object was not straightforward; left, right, up and down seemed to have been replaced by some strange new geometry that seemed to have been contrived specially to ensure that the observer always ended up in an unbelievably contorted position. And how on earth was I supposed to move the telescope to objects near to the pole? (The school's German equatorial mounting seemed to have been specifically designed to prevent this.)

It isn't easy – which is, I guess, why many visual observers prefer alt–azimuth mountings!

Commercial Telescope Mounting Systems

As far as commercial telescopes are concerned, designs have tended to become fairly standardised; either because there is an obvious "best" engineering solution or because there is a best economic solution, or both. It is the same with automobiles.

Schmidt–Cassegrains used to be manufactured with a single standard design equatorial fork mounting that included a "wedge" that enabled the fork to be tilted at the right angle (parallel to the axis of the Earth) in most latitudes. But since Meade introduced their LX200 range of telescopes in 1992 and computer-controlled equatorial tracking became affordable, the standard Schmidt–Cassegrain has become available in two mounting styles: alt–azimuth and the traditional equatorial wedge with fork.

A third variant has proved quite popular as well: Celestron now offer their Schmidt–Cassegrains on a "German" equatorial mounting. In a German equatorial mounting (shown in Figure 6.1) the telescope is mounted at one end of the declination axis with a set of counterweights (to balance the telescope about the polar axis) sitting at the other end.

Let's have a closer look at these mount options.

Fork Mountings

Figure 6.1. The German equatorial head of the author's 36 cm reflector.

The standard "fork and wedge" Schmidt–Cassegrain mounting is shown in Figure 6.2 (*overleaf*). The wedge tilts the fork over such that the fork arms point towards the pole star. The standard Meade Schmidt–Cassegrain wedge is adjustable for latitude and is optimised for best performance at a latitude of around 45°, with practical

Figure 6.2.
Standard "fork and wedge" SCT mount.

limits between 25° and 65°. It therefore caters for the vast majority of inhabitants of the USA and Europe. Meade's 180 mm aperture LX200 Maksutov and 300 mm aperture LX200 Schmidt–Cassegrain (both of which have relatively long tubes) are unable to completely clear the fork base for observer latitudes below 45°, and for observers below 45N the view of the southern horizon will start to become restricted.

For the majority of observers, this is of no concern. The standard Schmidt–Cassegrain fork mounting is extremely compact and lends itself well to fast slewing, where inertia and leverage considerations are all-important.

The only major drawback of the standard fork and equatorial wedge, when used with a Schmidt–Cassegrain, is the unexpected inaccessibility of the polar regions: when the telescope is pointed near to the pole the eyepiece disappears between, or close to, the fork tines!

Although equatorial fork mountings have become the standard for Schmidt–Cassegrains, they are rarely seen on any other form of commercial telescope. However, amateur telescope makers quite often build fork mountings for home-made Newtonian reflectors.

At the age of fifteen I built my own 220 mm Newtonian (including grinding and polishing the mirror), and built a hefty equatorial fork mounting for it (Figure 6.3).

Undoubtedly the size and weight of an equatorial fork mounting, when considered for medium-to-large Newtonians, has excluded it from most manufacturers' considerations – despite its suitability from the amateur telescope makers-viewpoint.

The mechanical design of commercial Schmidt–Cassegrain fork mounts can sometimes leave a lot to be desired. Worm, wheel, motors and gears are small and, to be fair, have to be built down to a price – one of the less fortunate effects of intense competition between major manufacturers. However, many of the mechanical inadequacies can be compensated for by sophisticated electronics and software design which allow

Figure 6.3. The author, aged 15 (1973), with his home-made 22 cm Newtonian.

Periodic Error Correction to be employed, along with CCD autoguiders to monitor and correct drive errors.

Overall the whole "Schmidt–Cassegrain telescope + computer-controlled fork mount" package is highly attractive. These "smart" features are also available on the German equatorial variants.

German Equatorial Mountings

Both my main Newtonian telescopes (0.36 m f/5 and 0.49 m f/4.5) are mounted on German equatorial mountings.

The German equatorial mounting is an elegant-looking design which has a number of advantages. First, the telescope is slung to one side of the supporting tripod or plinth, which means that the observer can get to the telescope eyepiece with less chance of tripping over the mounting in the dark. Second, as far as refractors, Cassegrains or Schmidt–Cassegrains are concerned, the eyepiece end of the telescope never becomes entangled with the mounting, so the whole sky is accessible.

An additional advantage of having a declination axis with counterweights on one end and the telescope on the other is that if the observing position is a bit inconvenient, the whole assembly can be rotated around the polar axis and the telescope swung around the declination axis and you have a second choice of position.

There is something else, too, rarely mentioned in other books. On my 0.36 m Newtonian the telescope tracking improves noticeably if the telescope is *over* balanced such that the heaviest component (telescope or counterweight) is on the West side. This serves to "take up" any residual slop in bearings and gears. Many astronomers learn that their telescopes have tracking idiosyncrasies of this sort. Adjusting the balance of a German equatorial mounting is easy: you just add or subtract lead counterweights, or move them along the support bar.

At this point, I will tell a cautionary tale. When I mounted my 0.36 m Newtonian/Cassegrain on a concrete plinth in 1980 I made sure that the concrete plinth was suitably massive – around 50 cm in diameter in fact. I then discovered that the mirror end of the telescope tube would not always clear the side of the plinth! Luckily the problem was not too severe and even in the worst case (viewing an object at the zenith) my blind spot was around forty minutes of time. It constantly surprises me how many times I have wanted to observe or image an object virtually overhead and have had to wait for forty minutes after the telescope has tracked into the west side of the plinth. Most irritating of all was when Supernova 1994 I in M51 was discovered. I was all set to take a good photograph – but the galaxy was at the zenith. After moving the Newtonian to the east side of the plinth I switched the drive off and waited for the galaxy to reappear; it reappeared just in time for a cloud to arrive, so my evening was a total waste of time! (The moral of this story is that it is essential to plan in detail the construction of a permanent mounting.)

Even if you buy a commercially-made telescope, you may want to mount it on your own custom-made plinth. The size and height of the plinth deserve careful thought; if you get it wrong, observer comfort may be compromised and other frustrations (see above!) may develop.

Celestron have, for many years, offered an alternative to the fork and wedge Schmidt–Cassegrain mounting, and Meade started offering a German Equatorial variant in 1997. Celestron's Schmidt–Cassegrain models are available on very elegant German equatorial mountings made by Losmandy, a company renowned for quality telescope mounts. Such mountings can easily be used for other purposes, e.g. for mounting a refractor, whereas the fork and wedge mounting is far less versatile.

The German equatorial, equatorial fork and alt–azimuth Dobsonian mountings probably account for 95% of all telescope mountings in amateur hands, but there are other variants.

The Split-Ring Equatorial

One of the most innovative commercial mountings developed in recent years is the mounting of the "Next Generation Telescope" (NGT) developed by JMI, a US company based in Colorado. Their "split-ring" equatorial is a horse-shoe type mounting, not dissimilar to that used on the Palomar 200 inch (5 metre) reflector. The horse-shoe is, effectively, a curved fork mount with the advantage that the horse-shoe outer ring (formed by metal tubing) can simulate a giant worm-wheel, except with a friction roller drive, not a worm gear, providing the tracking. JMI use this mounting to support their 320 mm and 450 mm Newtonians.

Commercial Drive Systems

So what options are available to the amateur who wishes simply to buy a drive system but not a complete telescope? And are computer-controlled facilities available with custom drives?

As you might expect, *any* option is available at a price – but quality, independent drive systems often cost as much as a complete Schmidt–Cassegrain telescope system.

Commercial, independent drive systems are – not surprisingly – centred around German equatorial mountings. A fork mounting design requires some pre-knowledge of the tube diameter and so is not particularly versatile! The leading suppliers of quality German equatorial mountings are the Japanese companies Vixen and Takahashi and the US companies Losmandy, Astrophysics and Byers. All five companies market a range of German equatorial heads to cater for a variety of instrument loads and tracking accuracies.

The smaller mountings are designed to carry small telescopes and have tracking accuracies suitable for 50–135 mm focal length unguided camera lens work (i.e. sub arc-minute periodic errors of the worm and wheel systems). The largest mountings cater for Newtonians up to 30 cm aperture or Schmidt– Cassegrains up to 35 cm aperture. These larger mountings typically have periodic errors of 10″ or so and often feature CCD autoguider input sockets and even Periodic Error Correction (PEC). As a rule-of-thumb, the average commercial telescope drive (without PEC and CCD autoguiding features) will not be good enough to allow high-quality, long, unguided exposures with focal lengths of more than 400 mm: star images will be small trails, not pinpoint dots.

I already mentioned (Chapter 2) that the Vixen GP Polaris head is a popular and inexpensive German equatorial mounting, especially in the UK. With the optional "Sky Sensor 2000" accessory this mounting gains fast slew, "GO TO" and PEC features. The larger Vixen GP "DX" mounting will support Newtonians up to 30 cm aperture and is available as an independent purchase, i.e. with no telescope attached. In the UK, Vixen mountings are marketed by Orion Optics.

Takahashi market a wide range of high-quality equatorial mountings; recommended maximum payloads range from 5 to 50 kg and reliability and tracking accuracy are excellent. The variable star discoverer Mike Collins has used a Takahashi mounting for all of his 200 discoveries! Takahashi mountings are marketed by Texas Nautical Repair in the USA and by True Technology in the UK.

Losmandy are well known as the suppliers of German equatorial mountings for Celestron's Schmidt–Cassegrain range. Losmandy are part of Hollywood General Machining Inc., based in Los Angeles. As with Takahashi, a range of mountings is available, for payloads from 15 to 70 kg. Losmandy claim that their impressive mountings are the only production mountings in the world where all the components are individually machined. They also market a range of secondary systems to enable guide telescopes, Schmidt–Cameras and other equipment to be piggy-backed onto existing mountings/telescopes.

Astrophysics (the company) have already been mentioned in Chapter 2 as the manufacturers of high quality apochromats. Their mountings have always had impressive tracking accuracy but now include additional features such as a fast slewing mode and a "GO TO" feature. Not only this, but their latest advertisements boast a "voice activation" system! The user simply says "Find Neptune," or "Find M1", and the Astrophysics mount (Meade products are compatible too) slews to the target.

I'd have to see this feature to believe it (speech recognition is still in its infancy), but "voice activation" apart, Astrophysics standard mountings are of the very highest quality. Also in this league, and for longer than most of the competition, Byers mountings have long been recognised as the standard by which the other mountings were judged. These days the competition is fierce and sheer machining accuracy and mechanical precision are not the only considerations: software control is equally important. To this end, Byers have recently teamed up with Software Bisque, a major force in Astro-software, to produce the Paramount GT-1100 Robotic Telescope mount. The claimed capabilities of this mount are impressive indeed. The mount can apparently be programmed to carry out a full night's observing session and the first four units were, allegedly, sold to the US Air Force to locate and track spy satellites! The main drive gear is the acclaimed

280 mm Byers Research Grade RA gear and the system comes supplied with the Software Bisque Professional Astronomy Suite which includes *The Sky*, *CCDSoft* (Image Processing Software) and *T-Point* (Professional Precision Pointing software that compensates for pointing errors). An impressive specification but, at $10 000, well outside most amateur's budgets!

At the other end of the scale (and for those observers who want the simplest of equatorial mountings) I'll describe the hand-operated "Scotch mount" in the next chapter.

Between the simple Scotch mount and the smallest, high-quality Vixen and Takahashi mounts there are a number of inexpensive camera mountings which will suffice for simple, unguided, 50–100 mm focal length photography of the constellations. The *Pocono Mountain Optics Series II* German equatorial mounting fits into this category and a very favourable review of this inexpensive mount appeared in the March 1998 *Sky & Telescope* (pp. 55–57).

Unfortunately, however expensive your commercial mounting, without a CCD (or human) autoguider, prime focus telescopic photography/CCD imaging is going to yield disappointing results. The cheapest equatorial camera mountings typically have periodic errors of 1′ or so and, as already mentioned, the best equatorial telescope mountings have periodic errors of 5″–10″. My 30 cm Schmidt–Cassegrain had a periodic error of 30″ prior to PEC training and about 5″ afterwards. However, for CCD Deep Sky imaging, where a pixel might span 2″, 1″ tracking accuracy is highly desirable! The inability of commercial drives to track perfectly is not solely a consequence of the limits of precision machining, or even foreign particles on the worm and wheel. When arc-second tracking is required, other factors – such as atmospheric refraction and telescope flexure–become critically important. Of course, the visual observer's requirements are nowhere near as critical; for them, the only requirement is that the object should stay near the centre of the field. To achieve this level of tracking even the crudest commercial or home-made drive will suffice. Many visual observers will be happy with a manual slow motion control, for example a simple flexible cable attached to the RA worm.

Some of the major telescope manufacturers supply add-on electric drives for their cheapest refractors and reflectors; these drives will satisfy the visual observer, but not the discerning astrophotographer/CCD imager.

Home-Made Drives and Unusual Mountings

Apart, perhaps, from the NGT "horse-shoe" mounting, unusual and innovative mounting designs are not generally found in the commercial telescope arena, but rather in the hands of amateur telescope makers, many of whom get as much fun out of building telescopes as out of using them; in some cases, more!

This is quite an important point as amateur astronomy is, in fact, a highly diverse hobby, with its members often far removed in interests and background from the "stargazing" stereotype. As well as innovation in the mechanical side of the mounting, many amateurs are skilled in electronic design as well.

Electronic Drive Design

Making a high-quality home-made drive usually necessitates some knowledge of electronics. Suitable circuits often appear in the popular astronomical literature. Electronic drives can generally be separated into three distinct categories, namely: d.c. motors; synchronous a.c. motors and stepper motors.

In all these cases the motor drives the telescope worm shaft via considerable gearing. The motors are often small, but although telescopes are big and heavy they fortunately only need to rotate once per day (except when being slewed).

Direct current (d.c.) motors are very simple: you just apply a voltage to them, and the speed varies according to the voltage! However, making them rotate at the precise rate suitable for CCD imaging is virtually impossible without electronic servo control systems.

Surprisingly, all leading manufacturers of Schmidt–Cassegrains use d.c. motors in their main RA drives. However, with *positional encoders* on the motor shafts and PEC, the failings of d.c. motors can be overcome. The reason for using d.c. motors in these applications is that they work over a huge speed range and are cheap and powerful for their size. In the Meade LX200 models the motors can fast slew the telescope at between 1400 and 1900 times the standard sidereal tracking rate (hence the noise!), so d.c. motors are ideal. The sidereal motor rate is roughly 8 r.p.m. before the gear train. For the DIY drive constructor, shaft encoders and PEC are not an easy option, so d.c. motors can only be employed if a poor sidereal tracking rate is acceptable, e.g. if the telescope is being used solely for visual observing.

Synchronous alternative current (a.c.) motors have been the mainstay of telescope RA drives since the 1960s. A simple oscillator circuit will keep the motor frequency reasonably stable; a quartz crystal-controlled oscillator will guarantee precision tracking even in extremes of temperature. My 0.49 m Newtonian features a 45 cm diameter phosphor-bronze wheel and stainless steel worm. With the quartz-locked oscillator drive I can get away with 10-minute CCD exposures on objects at the zenith – but then 45 cm diameter RA worm-wheels are rare!

More and more amateurs are now using stepper motors to control their telescopes. The drive shaft of a stepper motor turns through a precise angle for each cycle of electric pulses. These pulses can be dispatched from custom digital circuitry or indirectly from the serial port of a personal computer. PC software can easily count the number of pulses and therefore deduce the position of the motor shaft and telescope to a high precision, because the motor rotates at a fixed and known amount for each pulse. A limitation of a stepper motor drive is that if pulses are applied to the motor too quickly, or if it is overloaded, it will lose steps and "slip" and the motor/telescope position will become uncertain.

To track at sidereal rates and retain the capability to image fine planetary detail, the smallest step sizes should correspond to substantially less than one arc-second on the sky. For a third of an arc-second step size, 45 steps/second are required to track at the sidereal rate. Unfortunately, even slewing at more than 50 times the sidereal rate with this step size will cause steps to be lost, so stepper motor drives are limited to maximum slew rates of a degree every few seconds, rather than the six or eight degrees per second enjoyed by commercial encoder regulated d.c. drives.

Many amateurs unfamiliar with electronics often "persuade" knowledgeable friends to design or build a circuit for them, which is one of the advantages of being a member of an astronomical club or society.

Another useful source of telescope-making ideas and tips is a small-circulation US magazine which I was recently introduced to: *Observatory Techniques* is published by Mike Otis, 1710 SE 16th Avenue, Aberdeen, SD 57401–7836, USA (E-Mail: otm@iw.net).

Stepper motor control kits can be purchased from electronic suppliers. These kits often provide interfacing to a PC serial port as part of the package and the software source code is sometimes provided for the home-programmer's benefit. Electronic innovation is relatively rare in the Amateur Telescope Making community. However, when it comes to making the DIY mounting hardware, amateurs are full of ideas!

Unusual Mountings

When amateurs consider building very large Newtonian telescopes they often find that traditional fork and German equatorial mountings, even large ones, are not substantial enough to support the weight. Often the most vulnerable part of the design is the single polar axis bearing, but if the polar axis can be supported at both ends, the mounting becomes far more stable.

One of the best examples of this type of mounting is the *modified English* or *cross-axis* mounting. In this design, a single, tapered beam (thickest in the middle where the declination axis supports the telescope) stretches from floor level to the top of a tall pier, to form the polar axis. The declination axis, as in the German equatorial design, must be counter-poised to balance the telescope.

A superb example of this type of mounting can be found at the Conder Brow observatory near Lancaster, England (see Figure 6.4, *overleaf*). Amateur astronomer Denis Buczynski (one of the UK's most colourful astronomical characters!) is the owner of this observatory which features a number of unusual instruments ranging from a massive 53 cm Newtonian on the modified English mounting, to an ultra-modern 33 cm f/3.5 computer-controlled Newtonian on a custom fork mount.

Advances in CCDs and computer control have meant that many ex-astrophotographic amateurs are now forsaking aperture for smaller, more compact, remote-controlled telescopes, although for visual observers,

"aperture fever" is still a strong influence. The move to smaller instruments, especially for those of us who use CCDs instead of film, has meant a shift away from large Newtonians requiring massive mountings.

Figure 6.4. Denis Buczynski with the cross-axis mounted 53 cm Newtonian at Conder Brow Observatory.

A large number of other equatorial mount variants exist but are so rarely seen, even in the hands of amateur telescope makers, so discussion of them in detail is not justified. However, a brief mention of some of the most interesting may be of interest. In addition to the *modified* English mounting mentioned above, there is, not surprisingly, a straightforward *English mounting*. This mounting has similarities to the modified English mounting in that the polar axis is supported at floor level and at the top of a tall pier; however, rather than one single tapered beam supporting the telescope and its counterweights, the telescope is slung inside a cradle and no counterweights are required. The only examples of English mounting which I have seen personally are in private observatories in Lancaster, designed for massive (e.g. 45 cm f/7) Newtonians at the observatories of Denis Buczynski and Glyn Marsh.

Even rarer, but with the enthusiastic support of some amateurs, are the forms of mounting which allow a fixed, or nearly fixed, eyepiece position. These mountings are often referred to as Coudé (or elbow) mountings, although this term does not cover every fixed eyepiece solution. The key to preventing eyepiece movement is to mount precision optical flats within the polar and declination axes of a telescope and mounting the eyepiece within the polar axis too. A typical Coudé solution is shown in Figure 6.5.

Many of these designs prevent the telescope from reaching the polar regions; they are also costly to manufacture and, in these days of compact Schmidt–Cassegrains (where the eyepiece movement is limited anyway) reserved for the connoisseur.

As well as the "Cambridge" and "Paris" Coudé designs, other fixed eyepiece mountings include the Springfield, Pasadena, Grubb, Ranyard, Pickering and Manent mountings. You may occasionally read of these designs in Amateur Telescope Making literature, but you will almost certainly never come across one of these mountings in a back-garden observatory. Indeed, the only situation where you may meet a fixed eyepiece telescope is in a public observatory, where children and adults alike need to have easy access to a convenient observing position.

Figure 6.5.
A typical Coudé mounting.

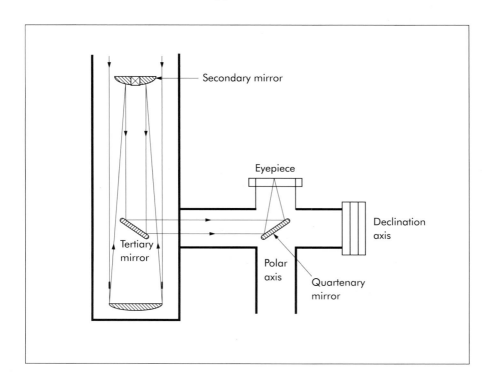

Poncet Platforms

On the subject of unusual mountings, there is a variant of the alt–azimuth mounting which enables an observer to take driven, long-exposure photographs using an alt–azimuth-mounted telescope. Known as a *Poncet platform* this is, essentially, an "equatorial platform" on which can be placed a Dobsonian or similar mounting (see Figure 6.6).

The platform – designed to be quite low, usually only a few inches high – sits on the ground. The upper surface can move – usually on castors – along small wedge-shaped inclined slopes. The shapes of the slopes are very carefully calculated so that when it moves, the upper part of the platform behaves as if it were a section of a very large cone, the axis of which is aligned with the polar axis of the Earth. In this way (and for a length of time depending on the design of the platform) the whole alt–azimuth mounting and telescope moves as if it were on an equatorial mount.

The ultimate Poncet platform may have been the one build by a remarkable British amateur, the late Jack Ells, who mounted his entire observing shed on the equatorial platform, so that he could observe in comfort –

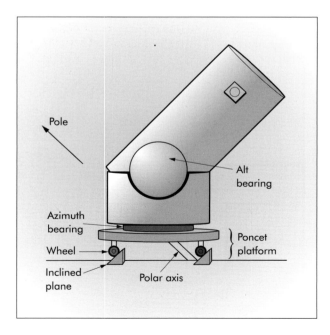

Figure 6.6.
Dobsonian on a Poncet mounting.

inside the heated shed – while it rode up inclined planes during the observing session!

Poncet platforms are manufactured commercially by a number of US companies. Although unorthodox, they combine the visual comfort of a Dobsonian with the tracking facility of an equatorial mounting.

Alt-azimuth Field De-Rotators

Merely driving an alt-azimuth telescope so that it "tracks" an object in the centre of the field of view, in an equatorial fashion, is not enough. The geometry is such that if the axis of rotation of a mounting is not parallel to the Earth's axis, the observed field of view will seem to rotate around the axis of the eyepiece.

The US telescope giant Meade are already offering so-called "field de-rotators" which allow their alt-azimuth Schmidt–Cassegrain telescopes to be used for imaging in alt-azimuth mode, without an equatorial wedge. These ingenious units fit into the drawtube of an alt-azimuth telescope and rotate any accessory (photographic camera or CCD camera) to compensate for the deficiencies of a non-equatorial mounting.

That concludes this look at telescopes and their mountings. But there are many accessories available to help the observer, and they warrant a separate chapter of their own. So let's now have a closer look at the gadgets and gizmos.

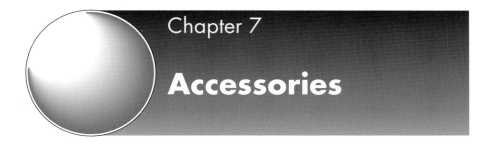

Chapter 7

Accessories

Finders

Possibly the most underrated astronomical accessory is the telescope "finder" (Figure 7.1). How many observers have been put off astronomy for life by their failure to locate the faint objects they want to look at?

The field of view of the human eye is very large, but the field of view of an astronomical telescope is very small: something in between is required to help point

Figure 7.1.
A standard 8 × 50 finder on a 30 cm LX200.

the telescope to a precision of one degree or less. A small, relatively wide-field telescope fixed to the main one was, and is, the obvious answer.

Unity-Power Finders

In recent years, so-called *unity-power finders* (that is, having no magnification) have become popular. These devices don't actually magnify the sky like a conventional finder scope, they project a red-dot or concentric rings against the night sky so that the telescope can be pointed to a spot pin-pointed by the naked eye.

Of course, they only *appear* to project the red-dot/circles on the sky; it's all done by virtual images partially reflected in small mirrors or glass panels; these devices simplify the classic observer's preliminary attempt to point the telescope, the so-called "squinting along the telescope tube" method. They are also easier to use than conventional magnifying finders as the observer does not need to contort himself to position his eye within a few millimetres of an eyepiece. Unity-power finders are ideal for wide-field, low-power telescopes and, with care, can be used to align the telescope on starfields to an accuracy of better than one degree. Their accuracy is limited by small parallax errors produced by movement of the observer's head.

Perhaps the best known commercial unity-power finder is the "Telrad" (Figure 7.2) which costs around $50 (1998). Telrad appeared in the 1970s, but in the inter-

Figure 7.2. The Telrad system.

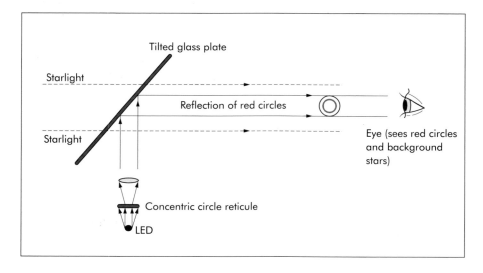

vening twenty plus years several Telrad look-alikes have also appeared. There is little to choose between the $50 or $60 Telrad clones. Some of them feature pulsed light sources, some project the image of an illuminated reticle, others project the image of an illuminated pinhole.

The only two commercial devices that are not really Telrad clones are the TeleVue "Starbeam" (at $230, strictly for the connoisseur) and the $23 "Star Site". The latter (yes it is a tenth of the cost of the former!) does not have a power source of its own, but uses a phospho-rescent ring which is energised for a few minutes at a time by shining a light at it – quite a novel approach.

One word of advice here. For those observers living in heavily light-polluted areas the advantages of a unity-power finder may be greatly reduced. Unity-power finders only work well when the observer can see plenty of naked-eye background stars to use as signposts. In light-polluted cities the naked-eye limit can be reduced to mag. 2 or 3, making it very hard to find enough stars to act as signposts.

Low-Power Finders

Most observers, even if they already have a unity-power finder and a dark site, will appreciate the facility of a second finder with a useful aperture. When search-ing for new comets or faint Deep Sky objects, it is essential that the observer is 100% confident that he is looking at the right field. Sometimes, even when you are *almost* sure that you have the telescope pointed at the correct region of the sky, the extra reassurance that a small finder can bring is immense. I have been in this situation many times myself when a moderately bright but very diffuse comet has eluded detection in the main telescope field. If the immediate narrow-field sur-rounding the comet is barren of stars, only a wide-field finder of, say, 40–60 mm aperture can resolve the issue.

On this theme, I have also found that printing out individual finder charts (from computer star maps) for each new object can be of immense help, provided that you choose the magnitude limit of the charts carefully. My own personal preference for my 36 and 49 cm Newtonians is a magnitude limit of 11. This is the rough limiting magnitude which I can see to with these instru-ments at low power, with only a minute or so of dark adaption, direct (as opposed to averted) vision and while switching between the eyepiece and map-reading.

What size objective should a finder have? Personally, I find 40 mm finders somewhat lacking, and I'd strongly recommend the use of a 50 mm, or even 60 mm finder with medium to large aperture telescopes. As with any accessory choice, cost will ultimately determine what is purchased, but the old adage "you get what you pay for" is still valid and relevant here.

Newcomers to observing may well be disappointed by the limiting magnitude when looking through a finder (or the main telescope for that matter!). Be aware that the limiting magnitudes given in popular astronomy books are the *absolute best* that an experienced observer can detect under excellent conditions, a far cry from what most of us experience. In theory, magnitude 6 stars are visible to the naked eye from a dark site: in practice, a city observer may well be challenged to detect stars of this magnitude *through the finder* against a light-polluted sky.

From my experience with large Newtonians and Schmidt-Cassegrains over the last seventeen years, I recommend a finder of at least one-fifth the aperture of the main instrument.

Having decided on *what,* consideration should be given as to *where* – that is, where the finder should be mounted and what orientation it should give. A finder that is a long way from the main telescope eyepiece is frustrating, as by the time you have moved from one eyepiece to the other, you have forgotten what the previous field looks like. And if the view through the main instrument is upside-down but the view through the finder is mirror-imaged or even erect, the mental confusion (especially in the cold, damp and dark) can be – well – irritating.

One solution that works well with Newtonians is to mount the inverting eyepiece of a "straight-through" finder close to the Newtonian eyepiece but on a rigid, tall stalk. This enables you to turn from the main eyepiece and use either eye to peer through the finder, which is especially useful if, like me, you always use the same eye for observing.

Guide Telescopes

If you want to use your telescope for long, guided photography of Deep Sky objects or comets you may need a further telescope, a *guidescope*, to help monitor and correct the tracking during the exposure.

If money is available, you might simply buy a CCD autoguider or CCD with built-in autoguider (these are covered in Chapter 8). But you may well have an old Newtonian or short-focus refractor incompatible with modern autoguider technology. In this case a separate guidescope will be essential. I even have a finder mounted on my guidescope!

Off-Axis Guiders

Another alternative to the guidescope is an *off-axis guider*. These useful devices were primarily developed as accessories for Schmidt–Cassegrain telescopes, but low-profile Newtonian off-axis guiders are also available from the US company Lumicon. Schmidt–Cassegrain telescopes have the advantage that a huge focusing range can be achieved (because focusing usually alters the inter-mirror distance, which in turn significantly alters the focal length). Thus with Schmidt–Cassegrain systems, accessories which take up a lot of space are not a problem because the focal plane can be moved in and out a long way. With Newtonians (and to a lesser extent, refractors), it is a problem.

Off-axis guiders use a tiny mirror or prism to deflect a small part of the light entering the focusing tube of the main telescope into a high-magnification eyepiece, usually equipped with an illuminated reticle. They are very effective.

I must say that I find it quite remarkable that so many sophisticated accessories are available to the amateur astronomer. In the past twenty years the amount of money amateurs invest in their hobby has increased dramatically. This, combined with electronics, CCDs and investments by companies such as Meade, has revolutionised amateur astronomy. Although it is a relatively rare pastime when compared with more common pursuits such as golf, fishing, or rock climbing, amateur astronomers now have the capability to take backyard images which would have been the envy of professional observatories only a few decades ago.

Horses for Courses

Serious observers usually make up their mind about what astronomical field to specialise in, and *then* decide what kind of telescope and accessories they need.

As with any purchase it is always a good idea to seek the opinion of others who may have more experience than yourself. The best people to consult will invariably be those who are known experts on a national basis, i.e. the most skilled observers, photographers and imagers. As with telescopes, many of the more expensive astronomical accessories are reviewed in the US magazines *Sky & Telescope* and *Astronomy*. In the UK, *Astronomy Now* has also carried out useful equipment reviews in recent years.

These reviews are well worth reading for an assessment of any potential acquisition. Because of the bewildering array of accessories available, the only way to tackle this subject is to split it into categories and deal with the more popular items in each category in some detail.

Photographic Equipment – Film versus CCD

You might imagine that the advent of the CCD revolution has closed the book on "wet" astrophotography. Nothing could be further from the truth!

CCDs will let you reach specific magnitudes with ten times less exposure than even hypersensitised film. In addition, image processing techniques enable the limiting magnitude of your instrument to be increased too; again, by a factor of ten, or even more. A 2.5 magnitude gain is impressive, but unfortunately it is not the only consideration.

Affordable CCD cameras typically have a few hundred pixels (picture elements – the smallest resolvable part of the image) along the width and the height of a CCD frame. If a CCD frame is printed in a magazine and enlarged to more than about 100 mm (four inches) along the longest edge the pixellation – "blockiness" – is immediately apparent. But an astrophotograph taken on fine-grain 35 mm monochrome film can be printed at a scale of at least three times this before the grain becomes objectionable.

And sheer sensitivity is not the only issue; when a bright comet comes along there is no shortage of photons around, but a large format photo of a "great" comet will look far more impressive and show much more of the structure than a pixellated CCD image.

The UK amateur astronomer Mike Collins has now discovered well over *two hundred* new variable stars in recent years using no more than a 135 mm camera lens and hypersensitised film (Figures 7.3 and 7.4, *overleaf*), providing a limiting magnitude of about 12.

The photographs of Comet Hyakutake shown in Figures 7.5 (*overleaf*) and 7.6 (*overleaf*) taken by me, Glyn Marsh and Nick James, were secured with no more than an 85 mm lens on a Vixen equatorial mounting. So don't underestimate the power of film!

Undriven Astrophotography

At its simplest, astrophotography consists of an undriven 35 mm camera on a fixed tripod. Such an arrangement can easily capture bright meteors during the major shower periods and is inexpensive and surprisingly rewarding. The simplest photograph I took of Comet Hale–Bopp (see Figure 7.7, *overleaf*) was one of the most attractive and also the one that was most popular with my friends who wanted copies!

With today's ultra-fast colour emulsions, a 50 mm lens at f/1.8 can secure pleasing sky photographs with

Figure 7.3. Mike Collins and his twin 135 mm Nova Patrol cameras.

Figure 7.4. Mike Collins and his blink comparator.

Figure 7.5. Comet Hyakutake: March 23rd 1996; 0248–0252 UT; 85 mm lens, f/1.8; 15° of tail; hypered Kodak 2415. From Tenerife. Bright star: Arcturus. [M. Mobberley, assisted by Nick James and Glyn Marsh.]

Figure 7.6. Comet Hyakutake in Ursa Major; March 25th 1996; 0109–0112 UT; 85 mm lens, f/1.2; from Tenerife; hypered Kodak 2415. [M. Mobberley, Nick James and Glyn Marsh.]

Figure 7.7. Comet Hale–Bopp; March 31st 1997; 2130 UT; 35-sec exposure; 50 mm lens f/1.8; Fuji SG 800. [M. Mobberley.]

a mere 30-second exposure. In such a short undriven exposure (and with a lens of fairly short focal length) the stars barely leave any trails on the film. With longer exposures the stars will leave trails, but 5-minute shots from dark country sites can record quite a few bright meteors during the major shower periods. Unfortunately, exposures of this length will invariably become badly fogged from town or city locations.

A variety of small equatorial mounting platforms is available for lens-based astrophotography. Vixen's excellent GP mounting has already been mentioned; Takahashi also market excellent small equatorial mountings.

Simple Tracking

Many amateurs have constructed their own "Haig", "Scotch Mount", or "Barn-door" mountings in which the polar axis is essentially a door hinge. The "Haig" mounting was popularised by G. Y. Haig in his paper "*A Stellar Spectrograph*" in the 1975 August *BAA Journal*. The three nicknames for this type of mounting have survived the intervening 23 years!

As the sky rotates, the "door" opens, at a rate of 1 degree every 4 minutes. Such a mounting can even be powered by the observer's hand turning a threaded rod, which opens the "door". For a 1 mm thread and a thread rotation of 1 r.p.m., the hinge-to-thread distance needs to be 229 mm (see the Figure 7.8, *opposite*). Winding a handle once every minute might seem primitive, but many amateurs have achieved stunning wide-field photographs with such simple equipment, especially when using fine-grain film.

Is there any kind of camera that is preferred for this kind of astrophotography? My advice here is: the simpler the equipment, the better. This is good news for those amateurs with a tight budget, as simple second-hand cameras are often far better than expensive, sophisticated ones. Modern 35 mm cameras are designed primarily for photography under a range of "normal" lighting conditions. Many have microprocessors which will use their own judgement as to whether the light levels are adequate for photography: when presented with an image of the night sky they will decide that photography is impossible under these conditions and refuse to let you go on!

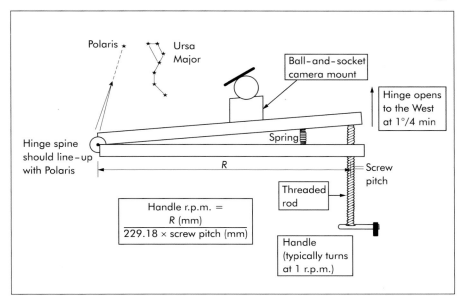

Polaris

Ursa Major

Ball-and-socket camera mount

Hinge opens to the West at 1°/4 min

Hinge spine should line-up with Polaris

Spring

R

Screw pitch

Threaded rod

$$\text{Handle r.p.m.} = \frac{R \text{ (mm)}}{229.18 \times \text{screw pitch (mm)}}$$

Handle (typically turns at 1 r.p.m.)

Figure 7.8.
Principle of the Scotch mount.

Another reason for considering second-hand camera bodies is that their low cost makes them more expendable. If you drop your second-hand $30 camera on the ground on a cold, dark night it is not a disaster. If you drop your $2000 camera it probably is!

One "old" camera that I can personally recommend is the Canon T-70. This camera went out of production in the mid 1980s but the second-hand market is still full of good examples. The Canon T-70 does have an electronic control system but – mercifully – in its manual mode it is astronomy-friendly in a number of important ways. Firstly, it uses two 1.5 V AA cells rather than the more traditional (and puny) "coin" cell batteries. A big battery means more exposures, especially in sub-zero temperatures which sap the strength of any type of cell. Secondly, the T-70 has a very frugal power consumption in manual mode. I have encountered SLR cameras which continuously meter the scene during a manual "brief time" exposure and waste energy in doing so; sometimes the batteries last for just one (astronomical) exposure! During the late 1980s I conducted an intensive Nova Patrol using a Canon T-70. After several years and thousands of 1-minute exposures I was still on the original set of AA batteries.

Finally, the Canon T-70 can be programmed to carry out exposures automatically and record the exposure data on the negative. This is achieved using the optional Canon "command back" (still available off-the-shelf in 1997) which clips to the back of the Canon T-70 (or T-80, or T-90). Command backs are available for a number of SLR cameras. They are expensive accessories but invaluable for meteor photography, fireball patrols or other events where "hands-off" operation is preferred (such as solar eclipses).

I must admit to having another reason for choosing a Canon T-70 for Nova patrol work. Some years ago in an article in *The Astronomer* magazine, the renowned amateur (and professional) astronomer Robert McNaught extolled the virtues of Canon's superb 85 mm f/1.2L Aspheric lens. Even fully open at f/1.2, this lens produces good star images. When stopped down to f/1.8 or so the images become excellent and the lens performs like a 50 mm aperture Schmidt camera! With my Canon T-70 and this 85 mm lens from a clear, dark, high-altitude site in Tenerife, remarkable results were obtained during comet Hyakutake's passage in 1996. A 3-minute exposure at f/1.2 on hypered Kodak 2415 (on a Vixen equatorial mounting) reached stars of magnitude 12.8 – not bad for an 85 mm lens (Figures 7.5 and 7.6)!

In the UK, 1-minute exposures at f/2 will yield a magnitude limit of about 11.0 with this lens. Unfortunately even a second-hand Canon 85 mm f/1.2L will set you back several hundred pounds in the UK.

Another excellent "old" camera is the Olympus OM-1 which features a lockable mirror (useful for avoiding vibration from "mirror slap" in short planetary or solar eclipse exposures) and an optional "clear" ground glass screen. Some modern cameras are worth a second look though, especially if you have deep pockets. The Olympus OM 4Ti and Nikon FM2N and F3 cameras are certainly worth considering. Both cameras have metering systems which work well for lunar exposures.

Dew

On those cold dark nights when you are sure that you *must* have been getting some good shots, make sure you periodically check (between exposures!) for the dreaded dew!

For meteor photographers or those using a piggyback camera (a camera fixed to the main telescope tube), a specially designed lens heater is a most valuable accessory. Unless your camera lens comes complete with a long dew cap ("lens hood" to a photographer), it will only take a few minutes to dew up on a typical damp evening.

A hairdryer would quickly solve the problem, *but is a safety hazard and must NEVER be used in damp conditions* (which is when you need it most). Low-voltage (12 V) hairdryer-like frost-removers for automobile windshields can be obtained.

Another solution is a lens heater made from thin, high-resistance wire threaded through a length of hook-and-loop ("Velcro") strip. The strip can be wrapped around the outside of the lens and joined where its two ends meet. By choosing a suitable length of wire, typically a couple of metres, and a suitable battery (I would recommend a rechargeable 12-volt battery with at least two amp-hours' capacity), an effective lens heater can be constructed. As a guide, the wire needs to get just perceptibly warm when connected to the battery to be effective; a few degrees Celsius of warming is more than enough to keep dew at bay. An alternative solution is to attach small resistors (values of 100 ohms or so seem to work best) around the periphery of the camera lens hood.

With either solution, the capacity of the battery will ultimately determine how much current can be passed through the wire/resistors for the duration of a long winter night. For preventing dew on Schmidt–Cassegrain corrector plates I would strongly recommend the Kendrick Dew Zapper system shown in Figure 7.9. This works on the same principle and is very effective, even on very damp nights.

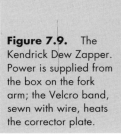

Figure 7.9. The Kendrick Dew Zapper. Power is supplied from the box on the fork arm; the Velcro band, sewn with wire, heats the corrector plate.

Film, Meteor-Photography Equipment and Wide-Field Camera Equipment

Modern colour emulsions, as I have already said, along with an inexpensive second-hand 35 mm SLR camera, can secure pleasing pictures of the night sky even with 30-second undriven exposures. At the time of writing (late 1997) I have two favourite emulsions which I would recommend for short exposure (by astronomical standards) work. These are Fuji's *Super G 800* and Kodak's *Ektar Pro-Gold 400*; both are colour negative films. The fine grain and colour balance of the Kodak film more than compensate for its slower speed.

Positive (colour-slide) films which are highly suitable for Astrophotography are Kodak *Elite Chrome 200* (formerly Kodak *Professional Ektachrome E200*) and Fuji *Astia 100* (= *Sensia II 100*). Do not be put off by the slow speeds of these films. The excellent reciprocity failure characteristics (see later in this chapter, under "Cold Cameras and Film Hypersensitising") and fine grain are more important than a fast ISO rating. In general, Kodak's films are best at recording red emission nebulae and Fuji's films are best at recording blue reflection nebulae.

A driven platform is unnecessary for meteor photography. The important requirement here is to take as many photographs as possible with as many cameras as possible! At the peak of the annual Perseid, Geminid or Quadrantid meteor showers, an observer may well see a meteor every minute or so. Unfortunately, photographic emulsions – even these days – are not as sensitive as the human eye.

Also, camera lenses do not in general have fields of view as wide as the effective "detection" field of the human eye. Exceptions are the so-called *fish-eye lenses* which can cover a 180° hemisphere, but with considerable distortion of the field and at a considerable cost.

The human eye will always spot far more meteors than will the camera lens.

The favoured focal length for meteor patrol work is 50 mm, the standard lens supplied with most non-zoom cameras. Standard 50 mm lenses usually have a maximum aperture of f/1.8 (or around 28 mm). Faster 50 mm lenses (f/1.4 and even f/1.2) will of course allow fainter meteors to be detected. Longer focal length lenses of larger aperture will also enable *fainter* meteors to be detected, but the narrower field of view will mean that fewer bright meteors are likely to cross the field.

Unfortunately it is a fact that the meteor capture rate (on film), even during the peaks of major showers, is disappointingly poor even with excellent black-and-white emulsions like HP5 or T-Max 400. I've typically watched meteor showers on crystal clear nights for hours on end with a camera pointed at the optimum position (50° altitude, 25° azimuth from the shower radiant) and have exposed several 35 mm rolls by the end of the night. After developing the films only one or two frames show any evidence of meteors, and even then they might actually be the tracks of artificial satellites! Cameras will only easily record meteors of first magnitude or brighter.

The only real solution to the problem of just a fraction of the sky being covered by a standard 50 mm lens, and the relative insensitivity of photographic emulsion, is the "sledgehammer approach" – just point as many 50 mm lenses as possible at the sky in the hope of ensuring that virtually all the bright meteors are captured. This, of course, is not an inexpensive solution, either in camera or film cost!

Die-hard meteor photographers use a standard platform or "meteor camera battery", as the assemblage of cameras is called. An example of such a camera platform is shown in Figures 7.10 (*overleaf*) and 7.11 (*overleaf*). The platform generally incorporates far more features than just "loads of cameras" and the design needs to be well thought out.

For a start, each camera needs a lockable cable release so that each shutter can be opened for minutes at a time. A command/data back on each camera would be the rich man's answer to this! Also, owing to various factors (atmospheric extinction versus the greater probability of more meteors when you look through more of the atmosphere, etc.) the optimum height to which each camera should be pointed to record sporadic meteors (not associated with major showers) is 50° above the horizon. For major meteor showers, the optimum position is generally regarded as 25° from the shower radiant.

Figure 7.10. A Lubitel meteor camera battery – note the propellor.

Figure 7.11. Lubitel camera battery – protective cover in place.

Bear in mind, if you envisage a long meteor session, that it may be necessary to change all the films in all the cameras during the night – because this is a hundred times more difficult to do in the dark with frozen fingers, some consideration as to how easily each camera can be removed and reloaded is essential.

Most of the meteor platforms I have seen have the individual cameras lying in angled wooden blocks, usually set at a about 50° for the reasons above. If more than six cameras are available, the remainder are often pointed at the zenith. A standard 35 mm camera fitted with a 50 mm lens will cover about $27° \times 40°$ of sky. Six such fields will be able to record most of the meteors passing through the sky at between about 37° and 63° altitude (approximately) – which leaves a 54° wide hole at the zenith. Spare cameras with 50 mm or shorter (wider-angle) lenses can patrol this.

Although this may seem to be a pretty "air-tight" meteor trap, in practice most meteors seem to have the knack of travelling along the edges of camera fields, between gaps in the fields or, in some cases, behind the only trees or buildings in the field. I have actually seen a photograph of a very bright meteor which chose to travel behind the leg of an electricity pylon as seen from the observer's station!

And don't forget that dew will form on any exposed glass surface given half a chance, so one of the anti-dew solutions needs to be applied to each camera lens.

Satellite Trails

I have already mentioned the problem of distinguishing between artificial satellite trails and meteors: the streaks they leave on film can look remarkably similar. Fortunately there is a solution, although it does mean building a piece of equipment which is not commercially available: the rotating shutter.

A rotating shutter looks like an aircraft propeller, positioned so as to cross the field of view of each lens on the meteor camera battery. The speed of the "propeller" and the number of "blades" can vary but – typically – each vane should interrupt the field for less than a thirtieth of a second. Because artificial satellites move relatively slowly, their photographic trails will not be broken by the rotating shutter. Meteors move very quickly so their trail will be broken.

If you know the precise speed of the rotating shutter, valuable information can be extracted about the speed of the meteor across the sky. It is therefore worth considering using a stepper motor to control the shutter speed. Stepper motor control circuitry can be designed (or purchased ready-made) to *precisely* set the rotation rate.

An additional effect of rotating shutters is that they prolong the maximum exposure time a little, as sky fogging is reduced by the blade interrupting the light from the background sky. Dew may also be partially held at bay by air movement caused by the rotating vanes.

But don't forget that changing film in a camera in the dark can be a tricky operation, and it becomes positively hazardous if you have to keep your knuckles from being whipped by a rotating shutter blade as well!

Medium-Format Cameras

Many meteor photographers use inexpensive or second-hand medium-format (6 × 6 cm) cameras on their camera platforms. At first this may seem like an expensive and pointless luxury – particularly in the cost of film – but the medium-format Russian "Lubitel" cameras cost as little as $30 brand new and offer a 44° square field at a focal length of 75 mm, thus giving a wider field of view and a higher resolution than the 35 mm camera with 50 mm focal length lens. Unfortunately the optical quality of the f/4.5 lens in the Lubitels is rather variable; there are good and bad examples. And the f/4.5 lens is much "slower" than the lenses fitted to even the cheapest 35 mm SLR cameras: but this does allow longer exposures before the background sky fogs the film (although you could always close the iris a bit on the 35 mm camera!).

Fish-Eye Lenses

The other option for covering large areas of sky has already been mentioned, using a "fish-eye" lens. A lens that covers the whole visible hemisphere of the night sky sounds like an ideal solution for meteor photographers, but in practice their main use is for recording fireballs (very bright meteors, far brighter than Venus and sometimes as bright as the full Moon).

High-quality fish-eye lenses are very expensive (well over $1500 in the US) and typically come in 8 or 16 mm focal lengths; one will fit the whole hemisphere of the sky within the standard 35 mm camera frame (36 × 24 mm), the other won't quite. As you can imagine, the image scale of an 8 or 16 mm lens is tiny: normal meteor trails

Figure 7.12.
Sporadic fireball captured by Henry Soper (Isle of Man) on 14th September 1991: 16 mm f/2.8 fisheye lens.

will cover only a small fraction of the field and considerable enlargement will be necessary to bring them out. The effective aperture of fish-eye lenses is also small, so only the very brightest meteors will be recorded, which is why fish-eye lenses are used for fireball patrols. A spectacular fish-eye fireball photo by Dr Henry Soper is shown in Figure 7.12.

Developing and Printing

The subject of developing and printing really warrants a separate book. However, I'll give a brief introduction here.

Most amateur astrophotographers soon lose patience with commercial processors of their black-and-white films. Commercial black-and-white processing is expensive, especially for large prints. In addition, the photo-shop or laboratory will almost certainly have no knowledge of astronomy and so is unlikely to centre the image correctly, or crop to the region of interest. Moreover, commercial printers will use a standard grade of paper (astronomers usually opt for "hard" – contrasty – grades, particularly for Deep Sky shots) and will charge enormous fees for custom printing (burning-in and dodging etc.).

The prolific astrophotographer will soon want his own darkroom equipment.

Developing Tanks

One of the cheapest essential pieces of equipment is the film *developing tank* (Figure 7.13). A developing tank is simply a light-proof cylinder containing a plastic or metal spiral and fitted with a lid that has a light-tight funnel. In the darkroom, the film is loaded onto the spiral (there are various ways of doing this, according to the type of developing tank: all of them require a certain amount of skill and practice). The spiral is then placed in the tank and the lid and funnel replaced. An air-tight stopper prevents the chemicals used to develop the film from spilling during the developing process. The rest of the job of developing the film can be done in daylight.

Figure 7.13. The Paterson developing tank – disassembled.

This simple and inexpensive item immediately frees the amateur from the clutches of commercial film developers. Even if you don't plan to make your own black-and-white prints, I strongly recommend processing your own films so you can simply select the best negatives for printing.

Transferring the exposed film from the camera to the developing tank doesn't have to be done in a dark room, or at night, because another useful accessory is available, namely the *film changing bag*. This is a light-tight bag with a double light-proof skin and sleeves through which the user's arms enter. This enables the film to be loaded into the tank in a fully-lit room without recourse to a wardrobe or darkroom. After a tiring night of astronomy it is tempting to go straight to bed and think about film processing the next day; a tired observer can easily make irreversible mistakes. However, loading last night's film the next day can prove difficult if there is no darkroom available. A film changing bag can be particularly useful in these circumstances and the author has found them invaluable for loading sunspot films in the daytime. One word of warning: in hot weather the user's hands can become very hot and sweaty inside a sealed changing bag, and films can be ruined by contact with sweaty fingers in the cramped space within the bag. Technical gloves, such as VT handling gloves, will solve this problem.

I have already mentioned that a tired observer can easily make silly mistakes when trying to develop films before retiring to bed. The late Harold Ridley, a renowned photographer of comets and meteors, told me a story once about the equally renowned late Reggie Waterfield, who was also a legendary British astrophotographer and astrometrist. In Reggie's time, comet photographers used large format glass plates for their photographs. The plates were held in special camera backs which allowed the plate holders or ground glass screens to be exchanged and large sliding shutters to be pulled back for the duration of the exposure. These camera backs can be quite a handful in the dark, but Reggie was an experienced comet photographer. After one superb clear night, Reggie guided a long-duration exposure of a bright comet and then took the camera assembly back into the darkroom. After extracting the glass plate, the usual careful developing, fixing and washing operation commenced. Reggie then turned on the dark room lights only to find he had developed, fixed and washed the ground glass

screen instead of the plate! The real plate was ruined –
it had been sitting on the darkroom sink when he
turned the lights on. Stories like this sound funny only
when they happen to others.

Developing Colour Film

I have found that using commercially available two-
chemical (developer and bleach-fix) kits for developing
colour film is also a relatively simple process. Without
any previous experience an amateur can produce
acceptable colour slides and prints with these inexpensive kits.

Don't do your own colour developing for the sake of
economy. In many cases it may well be more expensive
to develop your own colour negatives than to get them
done commercially, but the advantage is that you can,
within an hour of exposing a film, have the developed
negatives in your possession ready to select the best
frames for commercial printing. Colour work demands
a careful regulation of the developer temperature:
sloppy temperature control can lead to poor colour
balance problems.

My advice is: buy a book on the subject! Developing
and printing techniques are beyond the scope of this
book but can be of great interest to the amateur
astrophotographer. Having said this, a few words of
advice – mostly what I have learnt from my own disasters, may help.

There are a number of things which can go wrong
when films are being developed and even after development, when they are being inspected. Many people are
tempted to use developers and fixers to their ultimate
limit, a cheapskate policy which invites disaster.
Personally, I always throw developer away when it
has developed more than half of the maximum
number of films recommended on the bottle or tin. I
apply the same rules to the fixer. When an important
comet is coming along I never take any chances and
mix fresh chemicals a week or so before my first
planned exposures.

For black-and-white work I prefer to use Kodak
T-Max 400 or hypersensitised Kodak 2415. (I will talk
about hypersensitising later on.) The former film is far
grainier than the latter but avoids the use of complex
hypersensitising equipment. For lunar photography I

can recommend Ilford XP1 400 combined with grade 3 or 4 photographic paper. If you are prepared to spend long periods in the darkroom, burning-in and dodging lunar photographs, the combination of a low-contrast black-and-white film and a high-contrast paper can produce spectacular results. For developing Kodak 2415 and T-Max 400 I recommend using Kodak's D-19 developer, but a few cautionary words may be appropriate here.

Firstly, D-19 is normally supplied in powdered form; it is mixed with warm tap water to form the working solution. After mixing, the developer "matures" until it attains full strength after a few days. If you use D-19 within a few hours of mixing the solution you will be very disappointed, especially if you are trying to secure good images of faint Deep Sky images or comets!

Hypered Kodak 2415 is best developed over 4–6 minutes in D19 developer. T-Max 400 is often given a development time of around 4 minutes by amateurs using D-19, although reducing this to 3 or even 2 minutes produces a far less grainy result, with a tolerable loss in magnitude. The use of a wetting agent is a good idea, and also a squeegee for preventing drying stains ruining the negative. However, Kodak 2415 scratches easily, so use the squeegee gently and hang the drying negative up in a relatively clean environment. Alternatively, a final rinse in purified water will do just as well, and is safer.

Black-and-White Printing

If you decide to go the whole way and print your own films as well (strongly recommended) there are plenty of inexpensive enlargers to choose from, or even dark room kits containing everything for developing and printing your own films. After a few weeks of experimenting you will wonder why you ever let someone else do your black-and-white printing. But let's not pretend that the printing process is that easy. Developing films is a relatively painless process and the whole operation can be carried out in daylight. Darkroom work, on the other hand, is carried out in the subdued dull red environment of a darkroom. Enlargers, developing trays and drying prints take up a lot of space and necessitate the construction of a separate darkroom (unless the bathroom is modified to serve a dual purpose). Nevertheless, there is a great deal of satisfaction in pulling a good print out of the fixer and waiting for it to dry!

Photographic paper comes in various formats, finishes and grades but the type that amateur astronomers will probably prefer is called "multigrade". Black-and-white papers come in four standard grades. Grade 2 paper gives a normal photographic print; grade 1 gives a softer (less contrasty) print, devoid of true blacks; grade 3 gives a high-contrast print and grade 4 an even-higher-contrast print (see Figure 7.14)! In fact, some suppliers can also offer grade 5 or 6 paper, for the ultimate in "soot and whitewash" results. Black-and-white multigrade paper, as the name suggests, allows the user to choose the grade at which the photographic paper performs. This is achieved by filters, provided with each multigrade pack. The paper

Figure 7.14.
Comet Hale–Bopp photographed by the author with a Takahashi, 17th March 1997; E-160 Astrograph (160 mm f/3.3 and hypered Kodak 2415); 0400–0415 UT. Printed on grade 4 paper.

is sensitive to the colour of the light falling on it and by selecting the appropriate filters, the effective grade of the paper can be varied. In general, the heavier the filtration, the higher the grade. This means that high-contrast work will necessitate long enlarger exposures. The reason that amateur astronomers will prefer multi-grade paper is that it is often difficult to tell what the appropriate grade is for any given negative. For photographs of galaxies and comets, a grade between 4 and 5 is usually best. For planets like Jupiter, with subtle cloud details, grade 3 or 4 is optimum, but for lunar prints from Kodak 2415, grade 2 will probably be preferred. Essentially, multigrade paper gives the amateur the flexibility to experiment and produce the best print.

Cold Cameras and Film Hypersensitising

Ever since A.A. Hoag published articles on cold camera astrophotography in the *Publications of the Astronomical Society of the Pacific* in 1960 and 1961, amateur astronomers have experimented with various techniques to try to make their film more sensitive to light.

Photographic film suffers from an effect known as *reciprocity failure,* a term that needs some explanation. At normal camera shutter speeds of $1/1000^{th}$ to $1/30^{th}$ of a second, the required exposure obeys a simple reciprocal relation to the light level: all other things being equal, if you halve the light level, you can double the exposure time and obtain an identical photograph.

Unfortunately, if you reduce the light levels by a thousand-fold or a million-fold, as is the case when imaging faint stars and galaxies, the simple reciprocal relationship fails. At extremely low light levels, faint galaxies and stars will never be captured, however long the exposure. (The reason for this is that if too few photons per second strike the warm and energetic silver molecules, they will fail to change the atomic structure of the film and a silver grain will not be formed.)

Fortunately, there are ways of correcting this situation. In a *cold camera*, the lower temperature of the film means that the molecules are less energetic and so a much smaller percentage of the photons is lost during the exposure. In film *hypersensitising* ("hyper-ing"), the film is evacuated in a vacuum and then baked in hydrogen or "forming gas" (a mixture of hydrogen and nitrogen) prior to exposure. Any water vapour or oxygen in the film emulsion is eliminated and the film is lightly fogged. The absence of water vapour and oxygen in the film prevents the film from being badly fogged during the baking process, and pre-fogging increases the chance of photon capture by the emulsion. Ready-made silver grains are ready and waiting for real astronomical photons to join them and form an image.

In the 1980s, hypersensitised Kodak 2415 Technical Pan film became the film of choice for astrophotographers. This film has exceptionally fine grain, high contrast and is nominally rated around 32 ASA. This is far too slow to be useful for Deep Sky photography in its un-hypered state. However, the film responds very well to gas-hypering and the result is a virtually grain-less emulsion with a sensitivity far in excess of a standard fast film. The grainless nature of 2415 has another distinct advantage. Photographic grain is akin to visual "noise" and astrophotographers are looking for the maximum possible signal-to-noise ratio. When using telescopes of 1 or 2 metre focal length, the typical star sizes of faint stars captured on the film are of the order of 3″ in diameter.

Unfortunately, with 1 to 2 metre focal-length reflectors, this is similar in size to the grain on push-processed 400 or 1000 ASA films, so stars can be totally lost in the grain (the background "noise"). A hypered fine-grain emulsion with high contrast is the ultimate requirement for the astrophotographer and Kodak 2415 satisfies this requirement.

Some amateurs have built their own gas hypering facilities, but most will not be prepared to go to these lengths and many will have a distinct aversion to having high-pressure cylinders of hydrogen stored in their garages – few amateur astronomers would pass up on the chance to float around in Earth orbit, but preferably in a spaceship, with all limbs intact, and not surrounded by remnants of their workshop!

The US company Lumicon has, for many years, supplied gas hypering kits, comprising a small cylinder of forming gas and a simple vacuum system and film chamber. These work well and are recommended.

Some manufacturers will also sell you ready-hypered film, which is the easiest solution. Hypered film should be stored in a freezer when not in use, but a journey of a few days in the post will not spoil the film's sensitivity, especially if the film is well sealed and wrapped in foil. However, a succession of trips in and out of the freezer and exposure to the damp night air over a period of months will take its toll on the sensitivity and contrast of any hypered film.

Hypering kits can be used for hypering both black-and-white and colour films, although colour films may exhibit a colour shift when cooled or hypered. Kodak Ektar Pro-Gold 400 (known as Kodak Pro PPF 400 in the USA) and Fuji Super G800 Plus are ideally suited to Deep Sky use, in hypered or un-hypered states. Optimum hypering temperatures, pressures and duration vary from film to film but advice on these is supplied with each kit.

Film hypering has become less popular since the advent of the amateur CCD camera, which is at least ten times more sensitive than even Kodak 2415, but it is always worth keeping a few reels in stock for when the next big comet comes along!

The hypering facility built by Dr Glyn Marsh and pictured in Figure 7.15 shows his 200 mm diameter tank, the black hydrogen supply pipe and the vacuum pipe. A

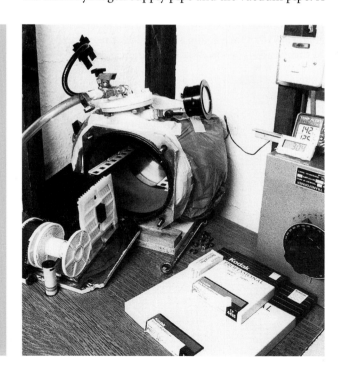

Figure 7.15. The hypering facility of Dr Glyn Marsh.

vacuum gauge is fitted to the tank and there are thermo-
meter modules for internal and external temperature
monitoring. A transformer unit powers trace heating
wrapped around the tank, under the insulation. Film is
vacuum dried and hypered while on the spool, NOT on
the cassette. The hypering cycle involves pre-heating the
tank, loading the film, evacuation for 1 hour at temperature, a brief purge with pure hydrogen, filling to *just*
above atmospheric pressure (only a few millimetres of
mercury above) and leaving for a pre-determined period
depending on the temperature set. For hypered 2415,
Dr Marsh usually uses 20 hours at an internal temperature of 33°C. But two hours at 45°C is often sufficient if
the film is used the same day. Dr Marsh emphasises that
cleanliness and dryness are the secrets to good hypering.
Film not used immediately should be sealed in aluminium foil and stored in a refrigerator or freezer. Such
film has a shelf life of at least a month and often longer,
especially if the film is not allowed to become damp on
repeated usage.

Cold cameras are rarely seen these days. In the 1980s,
film hypering became the preferred technique, largely
because of the hassle associated with using cold
cameras. A cold camera generally consists of a camera
body which can interface to a telescope and can also be
filled with dry-ice at a temperature of about –70°C
(Figure 7.16). Once again, a gas cylinder needs to be kept
in the garage (carbon dioxide though, not hydrogen).

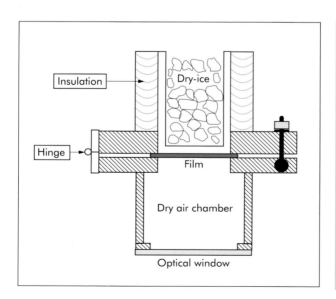

Figure 7.16. One
variant of cold camera
– the dry air cold
camera.

A transparent optical plug is necessary to insulate the interior of the camera from the warmer night air and the whole unit needs to be well insulated.

Unfortunately, a cold camera is nowhere near as versatile as a normal SLR. Most are simply receptacles for one sliver of film and a chunk of dry ice! Alternative focusing methods need to be devised, as there is no flip-mirror, pentaprism or ground glass screen, and the whole camera needs to be taken apart when another sliver of film is inserted. Few amateurs want to add to the work involved in an astrophotography session by fiddling about with a bulky gas cylinder and freezing cold dry-ice (which can cause serious skin burns) in the middle of the night.

The use of cold cameras peaked in the 1960s and 1970s when the spectacular results of Evered Kreimer were published in *Sky & Telescope* and in the *Messier Album*. Using a 310 mm f/7 Newtonian, Kreimer was able to photograph the Messier objects in relatively short exposures from Arizona, down to a limiting magnitude of 19 or so – a remarkable achievement in those days.

To my knowledge there have only ever been two manufacturers of cold cameras: Celestron International and Jack Newton. Celestron's cold camera was introduced in 1974; a few years before Provin & Wallis had introduced the readers of *Sky & Telescope* to the technique. The camera has been out of production for many years. The renowned Canadian astrophotographer Jack Newton has, for many years, manufactured his own cold cameras for discerning specialists; as far as I know he still does.

CCD technology has had a big impact on film hypering and cold cameras, making the former less popular and the latter pretty well obsolete. CCD technology is convenient and powerful, but hypered 2415 is still king when you want large and spectacular astrophotographs: keep some in your fridge!

Camera Interfaces

Photography through the telescope (as opposed to piggy-back photography with a camera "riding" the telescope tube) requires some means of interfacing the camera to the telescope. Most telescope suppliers stock a range of camera adapters to join most camera bodies (bayonet or screw thread) to 1.25-inch (31.7 mm) or 2-inch (50.8 mm) drawtubes. Some of these adapters allow an additional section to be inserted between the drawtube and the camera to permit *eyepiece projection* for lunar and planetary photography.

Fundamental to the design of these adapters is the so-called "T" mount, an international standard adapter which consists of the camera's own interface system (bayonet or screw thread) at one end and a 42 mm female thread at the other. Accessories such as universal camera lenses, microscopes or telescope drawtubes can be interfaced to the 42 mm female thread. All reputable telescope manufacturers supply accessories to enable their telescopes to mate with a female "T" thread; predictably these accessories are called "T" adapters.

A 42 mm thread can also be found if you unscrew the $1\frac{1}{4}$-inch (31.7 mm) end of most eyepiece projection units, so they can be inserted between a "T" adapter and the camera's "T" ring. Extension rings are also available with this thread in case extra eyepiece projection distances are required.

Unfortunately, there is a trap for the unwary. There is *another* 42 mm system, known as the "Pentax" or "M42" system. This also features a 42 mm interface which originally was used to attach the lenses to Pentax cameras. The pitch (or separation between the grooves of the thread) of the 42 mm "T" system is 0.75 mm; it is 1 mm on the M42/Pentax system. In practice, M42/Pentax threads will mate with "T" threads, but not in a totally satisfactory manner. Typically they can only be twisted together for less than one single turn, making the joint rather unreliable and prone to jamming.

My Starlight Xpress CCD camera heads have a female M42/Pentax thread and were hung precariously from my male T-threaded drawtube until I managed to acquire a Pentax to "T" adapter (Hanimex type 136915). Such adapters are available from most photographic shops and are inexpensive. Of course, if your CCD camera head does feature an M42/Pentax thread it makes it very easy to use cheap second-hand Pentax threaded lenses with the camera, if you need a wide field of view.

Focusers

Deep Sky and comet photographers in particular should, when choosing a focuser/camera interface, pay special attention to the accessibility of the focus and the vignetting of the light cone. For optimum performance at the Newtonian focus, a low-profile helical or electrical focuser (*"moto-focus"*) is recommended. This is simply a motorised focuser which can be controlled remotely via a cable. Meade supply electric focusers to fit their own LX200 series of telescopes, and these can be operated through the telescope remote-control panel. Electric focusers are much smoother to operate than the standard rack-and-pinion types and have the big advantage that they do not cause any vibration of the telescope when you adjust the focus.

For the CCD user an electric focuser has the added benefit of enabling focusing of the telescope to take place while the observer is looking at the monitor screen – indoors.

Of course, commercial Newtonian reflectors come supplied with some form of focuser; however many of these are unsuitable for astrophotography, and the first step that an astrophotographer should take is often to install a more suitable model.

A *low-profile focuser* is important because it enables the camera to be placed as close as possible (on a Newtonian) to the secondary mirror (see Figure 7.17). As well as increasing the probability of camera focus being achieved at all (with a "deep" drawtube, the film plane can sometimes be further away from the telescope

Figure 7.17.
A low-profile helical focuser + camera + magnifier.

tube than the focal plane of the primary mirror), this increases the area of the unvignetted field of view. Vignetting is one of the most irritating problems in Deep Sky photography and can easily ruin an otherwise excellent photo (see Figure 7.18). To minimise the degree of vignetting, try to buy a low-profile focuser with a draw tube of *at least* 2 inches in diameter.

If photography is your main aim, a standard $1\frac{1}{4}$-inch focuser falls far short of the ideal and may well halve the light grasp away from the centre of the negative. (Incidentally, the camera body itself will cause some vignetting unless your telescope is slower than about f/6 or f/7, in which case the effects will be tolerable.) If you intend doing a lot of visual work as well as photography, then you may have to compromise in your choice of focuser, i.e. low profile or not, for a Newtonian. However, even if visual work is your main interest, a 2-inch (50.8 mm) focuser must be regarded as essential these days, if only to exploit the range of superb wide-field eyepieces available.

The position of the Newtonian drawtube varies considerably from the photographic to the visual position, and the option of inserting filters, off-axis guiders and other auxiliary equipment has to be taken into consideration. My advice to the potential purchaser of a focuser is to consider the present and future applications of the telescope carefully before a decision is made.

Figure 7.18. One source of a vignetted image is often the camera drawtube.

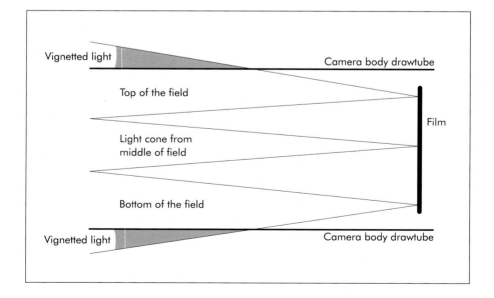

Vignetted light

Camera body drawtube

Top of the field

Film

Light cone from middle of field

Bottom of the field

Vignetted light

Camera body drawtube

Amateur telescope makers may find that in a Newtonian telescope, the choice of focuser influences the size of the secondary mirror. To determine the optimum focuser and secondary mirror size for your system, a simple ray diagram, showing the light cone, needs to be drawn. If prime-focus photography is contemplated (the option should always be considered, even by the hardened visual observer!), then the focal position should be at least 65 mm out from the mouth of the focuser when fully in; that is, flat against the tube. If an off-axis guider is to be used, then 80 mm is a better distance.

Unfortunately, the calculations don't end there. Other accessories may be placed in the light path before the camera, necessitating a point of focus even further out from the telescope tube. Filters and coma correctors are the worst offenders, and may result in your focal plane being as much as 120 mm out from the top of the focuser housing. While this in itself is not detrimental, problems then arise for the visual observer.

Most eyepieces focus when the focal plane is just outside, or just inside, the field lens. Thus, in a system like the one just described, the rack-mount/focuser tube may need to extend as much as 150 mm to hold the eyepiece in position. Not only does this result in an unwieldy system, but it necessitates a drawtube of similar length, which may vignette the light cone in fast optical systems. For example, to illuminate fully a mere 20 mm circle on film at f/5 requires the throat of the drawtube, 150 mm away, to be approximately $(150/5) + 20 = 50$ mm in diameter. This is a very borderline situation, even with a 50 mm diameter drawtube, when trying to avoid vignetted images.

An allied problem I have encountered is that some helical focusers described as "two inch" are only this wide at the hole into which the drawtube fits. This is OK for photographic and visual use, with appropriate "T" and 31.7 mm adapters. However, it is totally useless for using 2-inch (50.8 mm) diameter wide-field eyepieces. If you have a focuser of this type then you will need to get a friend with a lathe to make you a 2-inch "external" to 2-inch "internal" adapter if you are contemplating purchasing a 2-inch eyepiece (Figure 7.19, *overleaf*).

The choice of focuser for a Newtonian telescope is a tricky one, particularly for the astrophotographer who may want to observe visually. If you try to cater for every accessory known to man (off-axis guiders, filters,

Figure 7.19.
A 2-inch to 2-inch converter.

coma correctors, etc.) then your drawtube will be too long for comfort in the visual position, causing potential vignetting problems and necessitating a larger secondary mirror.

Commercial Schmidt–Cassegrain Focusing

Most of the above problems are irrelevant where Schmidt–Cassegrains and Maksutov–Cassegrains are concerned, for two reasons. First, the major manufacturers of these telescopes have already come across all the problems, and they provide accessories for most eventualities.

Second, these telescopes focus by moving the *primary mirror*. The resultant change in focal point (position of the focal plane of the optical system) for a small movement of the primary mirror is substantial, so problems of access to the focal point are rare. The accessories stay put while the primary mirror – and the focal point – move.

Manual Guiding and Off-Axis Guiders

The drive mechanisms attached to the polar axes of amateur telescopes are fine for visual observers who just want to keep the object of interest in the field, but they fall short of perfection as far as photographic or CCD observers are concerned (see Chapter 6).

CCDs and photographic films, placed at the focal planes of even modest telescopes (1–2 metres focal length) are capable of resolving down to arc-seconds, but conventional telescope drives cannot track to that accuracy. (A simple calculation will show that the machining accuracy required, for a telescope with a small worm and wheel to track to this precision, is unachievable. Small worm and wheel mechanisms cannot be machined to sub-micron tolerances [a micron = one-thousandth of a millimetre] and remain affordable!)

Modern Schmidt–Cassegrains often feature periodic error correction – discussed previously – and CCD autoguiders and CCD imagers with built-in auto-guiders (see Chapter 8) can monitor the drift of a star in the telescope field and correct the telescope drive to recentre the field automatically.

However, not all observers want to use (or can afford) the high-tech approach, and astrophotographers who prefer to use film, with its wider field of view and in-expensive technology, may well opt to purchase an *off-axis guider* system.

An off-axis guider is a device which fits between the observer's camera and the telescope drawtube (Figure 7.20, *overleaf*). A tiny prism (or mirror) in the off-axis guider body, placed at the edge of the field of view, diverts a small amount of light out of the off-axis guider and into a guiding eyepiece. Because the prism is well in front of the film plane and at the very edge of the field, it leaves hardly a trace of a silhouette on the film. Off-axis guiders are usually designed so that the eyepiece can be swivelled into a reasonably convenient position while guiding. I say "reasonably" because – as any owner of a large Newtonian will tell you – guiding positions are seldom actually comfortable!

The advantage of an off-axis guider, when compared with a separate guidescope, is that the guiding eyepiece is literally connected to the camera and is studying a star at the edge of the camera field.

A separate guidescope, even one that is securely mounted to the main telescope, can flex away from the main telescope tube, and differential atmospheric refraction at low altitudes can cause stars in the field to appear to move even if the guide star does not. Intercepting the light heading for the film, only centimetres from the film plane, prevents these problems.

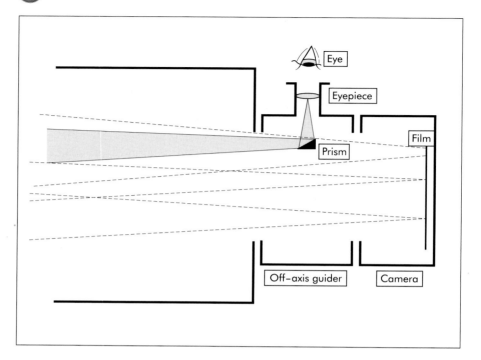

Eye

Eyepiece

Film

Prism

Off–axis guider

Camera

On the other hand, off-axis guiders do have disadvantages. Because light is collected from the edge of the field, where optical aberrations are at their worst, the guide stars will often appear highly distorted. And off-axis guiders may end up in inconvenient positions on large Newtonians and are far less versatile, for comet photography, than a separate guidescope. I'll talk more about this later.

A rigidly mounted, separate guide telescope will allow you to guide a Newtonian for 10 minutes or so without excessive flexure between guide scope and telescope. Ten minutes will take most f/5 Newtonian photographers near to their sky fog limit anyway.

However, if you really want to go for hour-long photographs of galaxies with fields of view no larger than half a degree or so then you will certainly be better off using an off-axis guider on a Schmidt–Cassegrain. During the 1980s a number of US amateurs secured stunning photographs of galaxies using off-axis guided f/10 SCTs (albeit with horrific exposure times of up to 3 hours in some cases!).

Commercial off-axis guiders are totally compatible with SCTs, and the slower optical system (typically f/10 or f/6.3) lends itself to long exposures, so that sky

Figure 7.20. The principle of the off-axis guider.

fogging takes longer. SCTs will also place the offset guiding eyepiece at a convenient height.

Guiding for Comet Photography

If there is one area of astrophotography that has not been commercially exploited, this must be it!

Comets move relative to the background stars and so have to be tracked by moving the telescope in a pre-determined direction at a pre-determined rate *relative to the stars*. A number of computer programs are available for determining the direction and rate of a comet's motion in arc-seconds per hour (3600 arc-seconds = one degree). I can recommend **eph.exe** by Nick James, available from Guy Hurst of *The Astronomer* magazine.

However, knowing the direction and rate a comet is moving are one thing: compensating for it quite something else! Most comets of mag. 10 and brighter move at rates between 100 and 500 arc-seconds per hour. Close Earth-approach comets, like Hyakutake or Iras–Araki–Alcock, can move at *thousands* of arc-seconds per hour (tens of degrees per day), a severe test for some sidereal drives!

Straight away, anyone wishing to take a photograph of a fast moving and bright comet is faced with a dilemma regarding the suitability of his astronomical equipment. Bright and close comets like Hyakutake are best photographed on hypersensitised Kodak 2415 or on fast colour film with exposure times of many minutes. During this time a nearby comet may move tens or even hundreds of arc-seconds. The intricate detail near the heads of these comets is drifting along at roughly the same angular rate as the comet's nucleus and fine details – mere arc-seconds across – can easily blur into invisibility if the comet's motion is not allowed for.

How accurately should one track a comet to capture the fine details which the telescope or lens is capable of resolving? A rule-of-thumb I have used in these case is to divide 6000 by the focal length of the lens in millimetres, to give the guiding accuracy required. For example, imagine you are using a 600 mm lens to photograph a bright comet which has a tail 2° or 3° long. The comet is moving at 1200″ (″ = arc-second) per hour. Using my "6000" rule, 6000/600 mm gives a guiding requirement of 10″ – so you have to guide to this accuracy or risk losing significant fine detail from the comet's head and tail. As the comet is moving at 1200″ per hour or 20″ per minute, it will drift through the 10″ tolerance boundary in only 30 seconds of time! Thus, if you rely on guiding your telescope by a nearby star, you will lose cometary detail after 30 seconds.

Fortunately, with very bright comets, like Hyakutake or Hale–Bopp, the nucleus of the comet is bright enough to guide on, using a separate guide scope of 60 mm or larger. Off-axis guiders are no help in this situation as the comet's head will need to be outside the field to guide on. (However, a few specialist companies manufacture "on-axis" guiders which can guide on an object in the field, albeit at the cost of a loss of light in the centre of the photographic field.)

For these bright comets, CCD autoguiders (see Chapter 8) can be used to autoguide on the comet's nucleus. And if a CCD is being used to take the picture, multiple short exposures can be taken and later stacked together to allow for the comet's motion. However, impressive comets are best imaged using film; CCDs available to amateurs just do not have enough pixels to match a quality photograph of a really bright comet. So what is the astrophotographer to do?

In days gone by, professional astronomers (and advanced amateurs) used *bifilar micrometers* in their guide telescopes. At regular intervals the observer would move the micrometer webs to recentre the guide star, thus moving the telescope in tiny increments away from the guide star. Bifilar micrometers are very expensive and so amateurs have sought other routes to enable them to track comets. You could for example easily obtain an illuminated reticle eyepiece with graduated scales (the Celestron Microguide eyepiece is an excellent example), and then simply use a position angle (PA) scale glued to the guide telescope drawtube to set the PA.

PA is measured from North (0°) through East (90°), South (180°) and West (270°). If you centre a star at zero degrees declination in your guide telescope and switch the drive off then the star will drift at 15 arcseconds for every second of time. This will enable you to calibrate the graduated scale *for that particular guidescope* in arc-seconds per division. You may need a Barlow lens to extend the focal length of your guide telescope if the reticle divisions are larger than the figure obtained by my "6000" rule. Typically, a guiding magnification of about one-tenth of the focal length in mm of the main telescope/camera lens is about right (e.g. 100× for a 1000 mm focal length astrophotography system). By turning the eyepiece so that the star drifts along the graduated scale with the telescope drive off, you can calibrate the East–West line. To guide on a comet you just need to turn your eyepiece through the required angle and calculate how often you need to reposition the guide star along the scale (such an arrangement is shown in Figure 5.8).

Remember that your guide telescope inverts the field! A useful expression to remember when you are moving up to the guiding eyepiece is: *move the star in the same direction as the comet is moving on the sky*. The human brain does not perform at its best in the dark, damp and cold, and memorising this simple saying can save many minutes of confusion at the eyepiece! Having said this, guiding on comets close to the pole has caused me considerable mental torture; I have yet to devise an expression for that particular nightmare.

Good news, really: this is a relatively inexpensive equipment solution to offset guiding. But what about the actual operation: surely, it can't be that simple? The answer, rather predictably, is *no*, it isn't that simple. Finding a suitably bright guide star to guide on can be a challenge, especially in barren star fields. This is where the comet photographer will appreciate a guidescope that can easily be moved with respect to the main telescope. Large but rigid adjustment rings are definitely required, and a separate finderscope on the guidescope can save many minutes of hassle.

Also, during guiding, the photographer will find that *total* concentration at the eyepiece and fast reactions on the telescope hand control are essential for a quality result. Even a few seconds spent looking away from the eyepiece at the stopwatch (which is invariably in a position suited only to a contortionist) can be disastrous. Personally, I prefer to offset the guide star without glancing at my watch. A cassette tape recorder can prove invaluable in this regard. By pre-recording audio messages at the required time interval (e.g. "move to division 1 ... move to division two") you can concentrate fully on guiding, without distraction. Admittedly, this method requires considerable preparation in advance (and can draw the occasional odd looks from any bystanders) but the pain is worth the gain (see Figure 7.21, *overleaf*). But please note, in cold

Figure 7.21.
Swift–Tuttle, November 13th 1992: 1816–1840 UT; 0.36 m f/5 Newtonian; hypered Kodak 2415. The exposure was meticulously and painfully offset-guided by the author for the 24-minute exposure!

weather, small tape recorder batteries have a habit of giving up the ghost!

Some amateurs have built automatic comet trackers which move the camera, or the guiding eyepiece, at the right rate and position angle to track a comet; if you have a workshop and a good knowledge of electronics/ stepper motors then this is definitely the way to go.

Time now for another rule-of-thumb. This one is applicable to comet photography. Sometimes, when a good comet is on the way (magnitude 6 or better), it is hard to guess what focal-length system to use to best capture the comet and its tail on film, from a really dark site. I devised the following formula for my own use – but if you like it, you can use it too! It is not based on any scientific principles and takes no account of viewing angle/elongation from the Sun. However, I find it helpful:

$$\text{Field of view in degrees required} = 10/(D \times R^{0.5} \times 10^{H_o/6})$$

where D = Earth–comet distance in AU; R = Sun–comet distance in AU; H_0 = comet absolute magnitude. So for Hale–Bopp on April 1st 1997 we have:

$$D = 1.36; R = 0.91; H_0 = -0.6 \text{ and the Field of}$$
$$\text{view} = 10/(1.36 \times 0.954 \times 0.794) = 9.7 \text{ degrees}$$

With 35 mm film, this corresponds to a focal length of 36 mm/tan 9.7 = 211 mm if the tail lies along the long axis of the film. Okay?

A Comet Marathon

To my mind, without a doubt the finest example of amateur comet photography ever made was the mammoth effort by Canadian amateur Peter Ceravolo and his colleagues in 1996. When Comet Hyakutake flew past the Earth in March of that year, Ceravolo and colleagues travelled to the Arizona desert, armed with a 440 mm focal length f/2.3 field-corrected Maksutov–Newtonian and dozens of reels of Fuji Super G 800 plus.

As the comet was passing so close to the Earth it was visible all night (unlike most comets which appear fairly close to the Sun), so photographic sessions were possible all night, during which time the comet's impressive ion tail changed noticeably. On their return to Canada, the team had no less than *900* negatives scanned and turned into a video – the first high-quality video of a comet moving against the background sky, with the tail flapping in the solar breeze – stunning. The video and a CD-ROM is available from Cyanogen Productions (see Appendix 1).

Coma Correctors

Coma is an undesirable optical aberration, especially severe in fast Newtonians. It manifests itself in the form of "tadpole" or "seagull" like star images at the edge of the field of view. Thus a coma corrector is a device which improves the quality of an image (not one which revives the observer from a trance-like state, as the name might suggest!).

Coma correctors are another class of accessory that take up space in the light path. They can also introduce vignetting if poorly designed. Large Newtonians of f/5 or slower will benefit least from their use, because the field of view will be too small to introduce coma, except at the edges. As most astrophotographers tend to enlarge their prime-focus images by about 10×, properly centred objects will not suffer adversely from coma. So a coma corrector is not essential unless you are using a telescope below 250 mm aperture and faster than f/5, or if you plan to use medium format (typically 6 × 6 cm) film. If you are building your own telescope of this sort of aperture then an f-ratio of 6 or 7 will give you definite advantages where coma and vignetting are concerned.

Commercial coma correctors are optimised for use with f/4.5 Newtonians. Certainly, the only situation for which I would consider buying a coma corrector would be for the amateur with a fast scope and medium format camera who is aiming for the maximum photographic field. With 35 mm film it is often the region of the field worst affected by coma that is also worst affected by vignetting from the camera body – quite fortuitous, in a perverse sort of way. So even if the coma is eliminated, the full width of the film cannot be used because of the illumination problems that vignetting produces.

An advantage of using a coma corrector with a Newtonian off-axis guider is that the distorted image of the star seen in the guiding eyepiece is significantly improved. Few amateurs that I know use coma correctors for photography, although a few enthuse about the improved visual experience when all the stars look sharp to the field edge. One of the most successful applications of a coma corrector that I know of is on the 33 cm f/3.0 automated telescope of Denis Buczynski, based at Conder Brow Observatory, Lancaster (Figure 7.22, *opposite*). This instrument was designed to illuminate an 8 mm square CCD with pixels covering 3″ or so. Unfortunately, at f/3.0, coma is severe and detectable even at the edges of an 8 mm CCD. However, Denis employed a Tele Vue Parracorr coma corrector, optimised for use on f/4.5 systems and, despite its use at f/3.0, coma was essentially eliminated.

In passing it should be mentioned that the Parracorr coma corrector increases the focal length of the telescope by 15%, so an f/3.0 system becomes f/3.5. Nevertheless, in this instance a coma corrector proved a good choice, even at such a fast f-ratio.

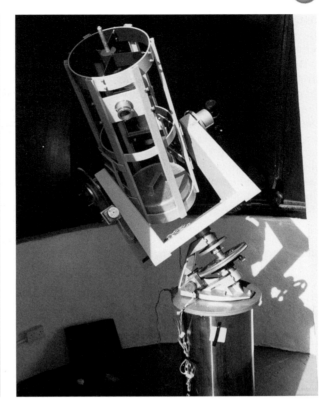

Figure 7.22. The 33 cm f/3.5 automated Newtonian at Condor Brow Observatory.

A Final Word on Focusers

If you can do without a coma corrector and an off-axis guider then you can have a very compact Newtonian focusing system: with the Newtonian drawtube fully in you are at the prime focus photography position – 80 mm further out and you can accommodate most eyepieces.

I strongly recommend helical or motorised focusers for astrophotography and CCD imaging, where focusing needs to be accurate, easy and, in the case of some CCD systems, remote controlled. The old-fashioned rack and pinion drawtube is fine for visual use, but precision is required for excellence in astrophotography and CCD imaging.

If you have a limited budget then my "bottom line" advice is: *don't* use a $1\frac{1}{2}$-inch (31.7 mm) focuser/rack mount for astrophotography; make sure you can reach the prime focus with a camera attached and the focuser fully in.

On the subject of accurate focusing I would also strongly recommend two other accessories. The first of these is a 3× magnifier for the SLR viewfinder; this will enable a sharp focus to be achieved on star images for Deep Sky and comet work. All major SLR manufacturers sell 3× magnifiers for their cameras. The second accessory I would recommend is a very fine ground glass screen. This will prove invaluable, especially for lunar and planetary work at long focal lengths. Most (but not all) SLR manufacturers can supply interchangeable screens.

Filters

The use of filters in amateur astronomy has increased dramatically in the last 20 years, owing to two main factors: the increasing encroachment of light-pollution and advances in filter fabrication technology.

Twenty years ago, an amateur with "a set of filters" was someone who had the right set of Wratten filters for planetary observation. Nowadays there are different filters for every conceivable type of visual and photographic astronomy. I will deal with each category in turn.

Planetary Filters

Much has been written on the use of Wratten filters, or their equivalent, for visual planetary observing and there is no doubt that different observers see different effects through the same filters. Venus, Mars, Jupiter and Saturn observations are all enhanced by the use of filters and there are four specific Wratten filters which are strongly recommended, as follows: Wratten No. 25 (red), Wratten No. 15 (yellow), Wratten No. 58 (green) and Wratten No. 47 (blue).

These filters are relatively inexpensive and gelatine squares of the appropriate type are easily obtained from any professional photographic supplier. Alternatively, telescope suppliers such as Meade offer their own filter sets which conveniently screw into the backs of eyepiece barrels. The use of filters will enable faint markings observed in white light to be enhanced and more easily recorded. For the observer with up to five favourite filters, the US company Lumicon markets a very useful "Multiple Filter Selector" for $129. This allows each filter to conveniently be "click-stopped" into position and the unit interfaces between a 1.25" drawtube and eyepiece.

The dedicated planetary observer may acquire considerably more Wratten filters as subtle changes in planetary atmospheres can be greatly enhanced by specific filters. The leading observers in the British Astronomical Association offer the following advice on filters for specific planets.

Venus

Almost any filter will reduce the dazzling brightness and aid better determination of the phase. Wratten 15 (yellow), 25 (red), 58 (green) and 47 (blue) are all worth experimenting with. Venus features subtle cloud markings but these are only detectable in the far-violet end of the spectrum. Glass absorbs UV radiation and so, for photography, a very long focal length Cassegrain with no intervening lenses is optimum. Fortunately, Schott manufacture special UV filters which transmit the region of interest (around 370 nm wavelength). The filters recommended are types UG2 and UV5.

Unfortunately, the human eye can see virtually no detail through these filters, but CCDs, although relatively insensitive in the UV, can record some cloud belts on Venus through the UG2 and UV5 filters.

Mars

Mars, observers benefit greatly from the use of coloured filters'. Subtle atmospheric and dust storm activity can be enhanced greatly by the use of the right filter. The Wratten 25 red filter and Wratten 47 blue-violet filter can be used to good effect when monitoring for dust storm activity using CCDs. The Wratten 25 or 15 (yellow) filters are also excellent visual filters for enhancing dark planetary markings. Enhancement of limb brightening and Martian cloud activity can be obtained by using Wratten 44A (blue), 58 (green) and 47 (blue-violet) filters.

Jupiter and Saturn

A survey of leading observers of the gas giants reveals a smaller number of filter users than the Mars' observers. Undoubtedly, Mars' observers, like Venus' observers, benefit from the use of any filter which will reduce the glare of the planet. Most observers of Jupiter and Saturn concentrate on using the Wratten 25 (red) and 47 (blue) filters.

The Moon

In addition to coloured filters, lunar observers may often benefit from the use of a polarising filter. The lunar surface, especially near full Moon, can be dazzlingly bright and, as with Polaroid sunglasses, glare from reflected sunlight can be reduced by a polarising filter. Variable polarising filter units can be obtained from Meade and Orion which allow transmission to be varied between 25% and 5%.

Deep Sky Filters

The problem of light-pollution has spawned a new breed of "interference" filters to specifically enhance the emissions of nebulae by suppressing the emissions from those irritating street lights.

Probably the best all-round "light-pollution rejection" filter is the Lumicon Deep Sky Filter, which blocks most sodium and mercury street-light emissions while letting through the vital emission lines of nebulae. The filter can be used visually or photographically and I have found that the combination of a Lumicon Deep Sky Filter and an ultra-fast film like T-Max 3200 is a powerful combination for nova patrolling from an urban site.

The filter does produce an attenuation of about a magnitude or so on photographic star images, necessitating longer exposures; but this is a small price to pay if you want to photograph nebulae from an urban location. The filter allows spectacular colour photographs of bright objects, such as the Orion Nebula, to be captured from city sites without any of the orange sky glow polluting the image.

Lumicon markets a whole range of other filters, most of which are extremely specialised and only for the Deep Sky purist. These more specialised filters are specifically designed to enhance single emission lines of hydrogen or oxygen, so as to enhance particular types of object such as supernova remnants or the Swan band emission line in gassy comets.

All of these "interference" filters utilise the so-called "Fabry–Perot" effect to achieve the narrow passband required. Readers may not be familiar with all the terminology employed here. The passband of a filter is simply the range of wavelengths that are allowed through the filter, that is, they are not attenuated. The stopband is the range of blocked wavelengths.

A narrow-band filter, quite often an interference filter, has a narrow passband and only allows a small part of the visual spectrum through. This may seem like a great disadvantage until one is told that most emission nebulae only emit their light in a very narrow band: eliminating everything else will dramatically darken the sky and enable the nebula to stand out.

Broadband filters, like the coloured Wratten filters, have a large passband and so the visual observer notices less of a drop in the brightness of the stars and a wider range of objects can be viewed. Obviously, the ultimate Deep Sky broadband filter would be one that eliminated every source of light pollution, but let through every astronomical emission line! Sadly, such a filter would be impossibly complex to make, but by choosing the right filter for the right type of object, dramatic contrast enhancements can be achieved.

Interference filters using the Fabry–Perot effect rely on multiple layers of dielectric materials to isolate the region of interest. Unfortunately, this type of filter only works well when the object is on-axis; objects at the field edge will be seen with less clarity. This can pose problems when using averted vision to utilise the most sensitive off-centre regions of the retina to view Deep Sky objects.

The Lumicon, Orion, Meade and Celestron Deep Sky filters are all general-purpose anti-light-pollution filters. Their passbands are designed to include all the major street-light emission bands. It is fortunate that most street-lights emit at discrete wavelengths and not across the whole spectrum! Unfortunately, High Pressure Sodium (HPS) and Incandescent street-lights emit over the whole spectrum and cannot easily be filtered. The easiest way to achieve a general-purpose

Mercury	405 nm	Pollution	Violet
Mercury	436 nm	Pollution	Blue
Hydrogen beta	486.1 nm	(Horsehead, California neb.)	Green
Oxygen III	495.9 nm & 500.7 nm	(Faint planetary–diffuse neb.)	Green
Mercury	546 nm	Pollution	Green/yellow
Neutral oxygen	558 nm	Skyglow "pollution"	Green/yellow
Sodium	570 nm	Pollution	Yellow
Mercury	575 nm	Pollution	Yellow
Sodium D	589 nm	Pollution	Yellow
Sodium	600 nm	Pollution	Orange
Mercury–sodium	617 nm	Pollution	Red
Hydrogen alpha	656.3 nm	Emission nebulae	Deep red

Deep Sky filter is to block all wavelengths between about 540 nm and 620 nm and below about 440 nm. In order of wavelength, the major street-light and nebular emission lines are detailed above and in Figure 7.23.

All of the Deep Sky filters available will darken a light-polluted sky while leaving the major emission nebulae unaffected.

As already mentioned, narrow-band filters for the Deep Sky purist are also available from Lumicon and Daystar. The Ultra-High Contrast (UHC) filters have a bandpass around 20 nm in width which include the single 486.1 nm Hydrogen beta line and the two Oxygen III lines at 495.9 nm and 500.7 nm. The Hydrogen alpha (H-α) emission line is blocked. These filters, as one would expect, provide ultra-high contrast views of

Figure 7.23. Filters, emissions and light-pollution.

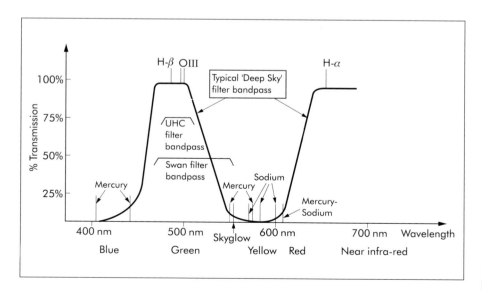

diffuse and planetary nebulae. The even narrower band OIII and Hydrogen beta (H-β) filters have a very limited range of uses. The bandpass is so narrow that background stars are severely attenuated. The OIII filter can be used on faint planetary nebulae or supernova remnants and the Hydrogen beta filter can be used on the Horsehead and California nebulae. For most observers, a UHC filter would be a far better investment than the individual OIII and H-β filters.

It should be borne in mind when purchasing a Deep Sky or narrow-band filter that even the relatively broad-band Deep Sky filters will result in at least a 1 magnitude loss for the visual observer and a doubling or trebling of photographic/CCD exposure times. The narrow-band filters will produce a very disappointing result when used with film or CCDs.

Comet Filters

The tails of comets are highly elusive features unless a great comet is being observed or the observer is privileged to be observing from a dark, high-altitude site. Dust tails can be spectacular (as in the case of Hale–Bopp) but they are rare and rather featureless, consisting merely of a swath of dusty material reflecting sunlight. However, ion tails, although more subtle, contain intricate detail as the ions are twisted by the interaction between the solar wind and the comet's tail.

The only truly effective filter for comet observing, especially gassy comets, is the "Swan band" filter. This filter has a passband covering the Swan bands of C_2 (roughly 470–550 nm) in the green region of the visual spectrum and will increase the contrast between a gassy comet with an ion tail and the sky background. Yet again, the US company Lumicon can supply such filters.

Photographic Filters

There are two photographic filters which are strongly recommended for photography with telephoto lenses. These are the H-α pass filters and the minus-violet filters; they are relatively inexpensive but can produce spectacular photographs.

The H-α pass filter totally blocks all light below 630 nm – that is, it lets only through red light and above. This means that virtually all forms of light-pollution are blocked but the H-α emission line (656 nm) is not. The only problem with using this filter is that it must be used with gas-hypered 2415 film, which is very sensitive to the red end of the spectrum. In addition, the lens focus should be set roughly midway between the normal and infra-red positions (consult your SLR manual for details). The contrast gains achieved with this filter and a camera lens are dramatic; from a city location you can easily photograph Barnard's loop in Orion and the complex filaments of gas around the Cygnus area. Even more spectacular results can be achieved using a Schmidt camera.

The minus-violet filter, primarily designed for sharpening colour telephoto images of star fields, does exactly what you would expect: it blocks the violet part of the spectrum, which gives rise to the bloated star images seen in many amateur photographs. This inexpensive filter will drastically sharpen colour telephoto photographs of the night sky, although you will need to increase your exposure times to achieve the same limiting magnitude.

Photometric Filters for CCDs (see also Chapter 8)

Professional astronomers measure magnitudes in accordance with internationally agreed standards: the sensitivity of their equipment is accurately defined at specific wavelengths. A series of passbands has been defined from the ultra-violet through to the infra-red region and the most widely used system for defining these bands is the Kron–Cousins UBVRI system. (U = Ultra-violet; B = Blue; V = Visual; R = Red; I = Infra-red).

Predictably, the V-band is the band of interest for most amateurs. It approximates the visual band of the human eye so that CCD magnitudes through a V-filter are directly comparable with visual magnitude estimates.

But the B- and R-bands are also of interest. Some variable stars vary considerably in the blue end of the spectrum, even more than they do in the visual band. CCD's are at their most sensitive in the R-band, so this region is also of interest. The I-band is of interest to specialist professional astronomers but the U-band is close to the limit of most CCDs' spectral range.

If the magnitude of a star is determined in the V- and B-passbands, the B–V (B minus V) colour index results. This colour index can tell us a lot about the star; it can also be used to calibrate a photometric CCD system.

When choosing filters for a CCD camera it is important to understand that the Kron–Cousins bandpass boundaries define an ideal system and it is not possible to perfectly match a given set of filters to a CCD. The CCD will have its own response at specific wavelengths and to derive the response of the system it is necessary to *convolve* the spectral response of the chosen filters with the spectral response of the CCD chip. By "convolve", we mean multiplying each point on the curve of the filters' spectral response, for every wavelength, with each point on the curve of the CCD's spectral response. This may sound impossible – but in practice, CCD and filter manufacturers provide spectral transmission data for their products at a large number of specific wavelengths; feeding this data into a math spreadsheet can easily simulate the response of the system.

Some authorities recommend different filter sets for different CCD cameras such that precise matching to the Kron–Cousins standard can be achieved. In practice this is not really necessary for V-band photometry to ±0.1 magnitude accuracy or for V and B photometry to ±0.05 magnitude accuracy. In the former case, the V-band lies well within the spectral range of all amateur CCD cameras and one filter should suffice for all commercial cameras at this required accuracy. In the latter case, using V and B filters and a calibration star field, the system can be accurately calibrated and measurements adjusted accordingly.

Most amateurs will, initially, be happy with simple V-band photometry; a few may, eventually aspire to progress to B–V and even R-band work.

The techniques used to calibrate and perfect such measurements are beyond the scope of this book but, fortunately, a number of excellent papers have appeared on the subject, in amateur astronomy magazines. The former US magazine, *CCD Astronomy*, published by Sky Publishing, carried excellent articles on this subject in the Fall 1995 and Spring 1996 issues, and an excellent paper by Richard Miles recently appeared in the April 1998 *Journal of the British Astronomical*

Association. This article recommended the Schott filters detailed below for photometry with amateur CCD cameras; for each passband, the thickness in mm of each filter is detailed:

U: 1 mm UG1 + 2 mm S8612
B: 1 mm GG385 + 1 mm BG39 + 1 mm BG12
V: 1 mm GG495 + 2 mm BG39
R: 1 mm OG570 + 2 mm KG3
I: 3 mm RG9
Clear: 3 mm WG280

Note that each filter set has the same total glass thickness so as to try to standardise loss of light by absorption and preserve the precise focus for each set. Also note the clear glass option which merely provides 3 mm of glass for the same reasons.

The glass filters can be cemented together to improve transmission but the choice of mounting is up to the individual. For super-accurate photometry, a precision filter-holder is essential, but the vast majority of amateurs in this field merely wish to obtain ±0.1 mag. precision on faint variable stars so they can "take-over" where the visual observers lose the object. My own filter-holder is far from precise (see Figure 7.24)!

Few visual observers make magnitude estimates below magnitude 15 and for those who do the accuracy

Figure 7.24. The author's MKI, rather crude, V-filter assembly.

is often unreliable. When novae and supernovae fade below this limit, it is time for the CCD photometrists to take over and provide accurate magnitude estimates of faint objects.

Although this may sound simple, it is worth remembering that a V filter will easily knock 1.5 magnitudes or more off your CCD camera's magnitude limit. When you add the fact that you need a good signal from the star (but not enough to take it beyond 50% saturation!) you need a surprisingly long exposure to get down to those 16th mag "V"-filtered stars that are beyond the visual observer! This is yet more proof of the formidable abilities of the human eye and brain combination, which can guesstimate a faint star's magnitude in an instant.

Schmidt Cameras and Astrographs

The subject of Schmidt cameras could occupy an entire chapter on its own. However, I have treated the Schmidt camera as an accessory simply because it can only be used for photography and is a rare possession, often piggy-backed onto the observer's main instrument.

The Schmidt camera was designed around 1930 by the introverted but brilliant optician Bernhard Schmidt (1879–1935) while he was working at the Bergedorf observatory. Remarkably, despite losing the lower part of his right arm while experimenting with explosives in 1894, Schmidt's excellence as an optical worker became legendary. In the 1920s, photographic emulsions were not as sensitive to light as the super-fast emulsions available nowadays, and hypersensitising was unheard of. In addition, the f-ratios available with lens-based astrographs were relatively slow, typically f/5 or f/6 with 20–30 cm refractors. What was needed was a super-fast astrograph which would enable nebulosity to be captured with ease and wide fields to be photographed in a single exposure. Schmidt's solution to the problem was the solution of a genius (see Figure 7.25, *overleaf*).

Firstly, he opted for a solution using a mirror and a lens, with the mirror having a spherical rather than a parabolic curve. A spherical mirror means that the focal plane is not only curved but also suffers from

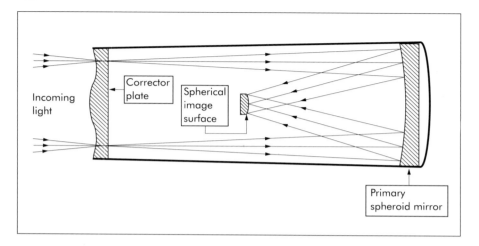

Figure 7.25. The Schmidt camera.

spherical aberration but coma, astigmatism and distortion – the limiting factors in a wide-field parabolic mirror design – are not present. Secondly, Schmidt introduced a thin aspheric correcting lens or plate at the other end of the telescope tube. Schmidt's "corrector plate" featured subtle curves which acted as a converging lens in the centre and a diverging lens in the outer zone, carefully calculated to eliminate spherical aberration. The subtle curves of the plate are weak enough that chromatic aberration is not sufficiently serious to be of concern. Another tribute to Schmidt's genius was the technique he used to produce the corrector plate: a combination of polishing and vacuum evacuation, using air pressure to distort the glass.

The diameter of the Schmidt "corrector plate" is the effective aperture of the Schmidt camera. This is usually slightly smaller than the diameter of the main mirror. The main mirror is "oversized" to avoid uneven field illumination which can occur if the primary mirror and aperture are the same. A similar "oversizing" is usually employed with Schmidt–Cassegrains. A set of three figures are often seen when reading about Schmidt cameras, e.g. 20/22/30; this means that the Schmidt camera has an aperture of 20 cm, a mirror diameter of 22 cm and a focal length of 30 cm.

The advantages of the Schmidt camera are awesome: a huge field with an ultra-fast (typically f/2) f-ratio and pin-point star images across the field.

The serious disadvantage of the Schmidt design, for amateur astronomers, is that the focal surface (where the film lies) has to be *curved*. The radius of curvature of the film is equal to half the radius of curvature (i.e. the focal length) of the spherical mirror. In fact the curve of the film is precisely concentric to the curve of the main mirror.

The curved focal surface and the fact that the focal surface is inside the tube of the Schmidt camera means that a standard 35 mm camera body cannot be used with the Schmidt, and the camera cannot be used visually. Single pieces of film, stretched around a curved surface, have to be used; this is not practical for most amateurs. Needless to say, preparing and loading a Schmidt camera with film is a delicate operation which must be carried out in total darkness and with the greatest care. The absence of an eyepiece means that the observer must have a separate telescope or guidescope aligned with the Schmidt camera and an exposure is literally started by removing the end cap on the telescope!

Other popular designs of fast astrographs are also seen in amateur hands, but none can quite match the extra-ordinary sharpness and photographic speed of the Schmidt camera.

The Japanese company Takahashi market an excellent series of fast astrographs called the *Epsilon* series. The author is the proud owner of an Epsilon 160 f/3.3 astrograph (see Figure 7.26). These astrographs can also be used visually and feature a hyperbolic primary mirror with a two-element field corrector. It transpires

Figure 7.26. The author's Epsilon 160 mm f/3.3 Astrograph.

that correcting aberrations such as coma in fast, wide-field Newtonians becomes considerably easier if the primary mirror is *hyperbolic*, not parabolic. Takahashi take advantage of this fact and the author has also found that the *Epsilon* design dispatches the light cone onto photographic film with the minimum of vignetting, even when used with a 35 mm camera body.

During the 1970s, in response to consumer demand, Celestron International brought out a range of three Schmidt cameras in 5.5-inch (140 mm), 8-inch (203 mm) and 14-inch (356 mm) apertures. The f-ratios were quoted as 1.65, 1.5 and 1.7 respectively. These Schmidt cameras became extremely popular with dedicated astrophotographers who tended to use the fastest colour films of the time (typically 400 ISO) with the cameras. Celestron even designed a special roll film holder for the cameras such that the cameras were not limited to single shot exposures. During the 1980s, amateurs found that *unhypered* Kodak 2415 could be successfully used in the cameras, by virtue of their remarkable f-ratio.

However, it was also during the 1980s that a number of other factors made astrophotographers rethink their choice of astrograph. As already mentioned, film hypering became highly popular during this period and the photographic speed of "fast" colour emulsions increased from 400 to 3200 ASA!

Companies such as Takahashi started producing astrographs of around f/3.5 which would take these fast films to the sky fog limit in ten or fifteen minutes, even from dark sites. The new astrographs could be used with conventional cameras and visually – a far more versatile product than the Schmidt camera. These days you rarely see amateur Schmidt camera photographs – at least until a good comet comes along!

Needless to say, the peak periods of Hyakutake (March 1996) and Hale–Bopp (March/April 1997) resulted in a host of Schmidt camera activity. The ultra-wide fields provided by these instruments are ideal for photography of such "Great Comets", and the fast speed and short exposures allow the effects of the comets' rapid motion to be less troublesome. In Germany, France, Austria and Switzerland some fine exponents of the art of Schmidt camera photography exist, and the names of Keller & Schmidbauer, Dahlmark, Jager and Klaus are synonymous with excellence in this field. Many of these amateurs use home-made Schmidt cameras.

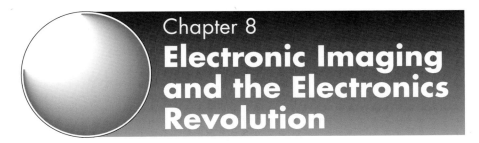

Chapter 8

Electronic Imaging and the Electronics Revolution

CCDs

The Basics

The same rule applies to CCDs as applies to photographic film: the bigger the area, and the higher the resolution, the better. Resolution in a CCD is determined by the number and size of light-sensitive elements contained along the length and the width of the device. These light-sensitive elements are actually Metal–Oxide–Semiconductor (MOS) capacitors which are transparent to light. In an amateur CCD camera there are, typically, a few hundred MOS capacitors along the length and the width of the CCD chip. Thus, there are over one hundred thousand MOS capacitors on every CCD. Each capacitor corresponds to a single pixel in the final image seen on the computer monitor.

When a photon reaches the semiconductor material there is a good probability that it will be converted into a tiny electrical charge. This charge can be stored in the capacitor and subsequent photons simply increase the charge stored. With over a hundred thousand MOS capacitors on each chip, each one collecting photons, the process can be likened to laying out over a hundred thousand plastic buckets on a football pitch on a rainy day; each raindrop being equivalent to a photon (Figure 8.1, *overleaf*). Indeed, like a plastic bucket, the capacitor-pixels can overflow if too many photons (raindrops) land on them and overflowed charge can spill into other adjacent pixels. This can happen when

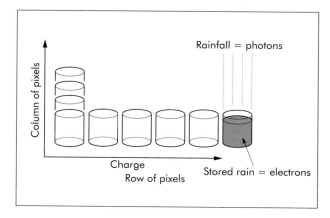

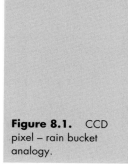

Figure 8.1. CCD pixel – rain bucket analogy.

bright stars are imaged unless so-called "anti-blooming gates" are present to deal with the overspill.

In the CCD these capacitor-pixels are incredibly small, typically 10 microns (or a hundredth of a milli-metre) across. Thus a CCD which has 400 × 400 pixels will have an imaging area of roughly 4 × 4 mm across. At the end of a CCD exposure the charge is transferred out of the capacitor-pixels by altering the polarity on the capacitors; the charge in each pixel is then transferred from an analogue charge to a voltage and thence into a digital number, ready for processing by a computer.

Two different basic types of CCD are used by amateur astronomers; the types define how the CCD image read-out is achieved. These two types are known as "frame-transfer" and "interline" (see Figures 8.2a, *opposite* and 8.2b, *overleaf*). In the "frame transfer" CCD there is an image zone and a memory zone on the CCD chip.

When the image is downloaded, the charge in the image zone is rapidly transferred to the memory zone. I stress *rapidly*, because if the process is not rapid, new photons arriving during a slow download could be smeared out over many pixels during the download process. Once the charge has been transferred to the memory zone, which is masked off from the light, a more leisurely transfer of the charge out of the CCD chip can take place. In practice, image smearing is only a problem when imaging bright objects (like the full Moon) when the light levels are high enough to enable significant numbers of photons to arrive during the download period.

In the interline CCD a memory zone is, in effect, created after each row of pixels, thus allowing a very fast readout and no image smearing, even on bright

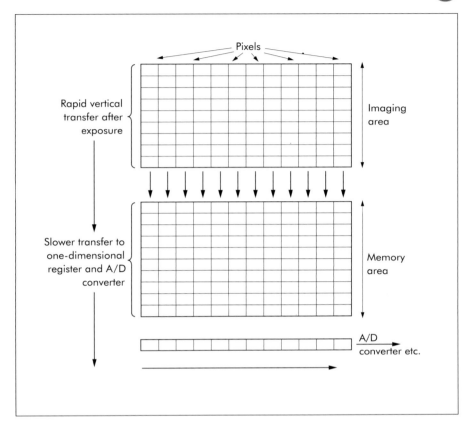

Pixels

Rapid vertical
transfer after
exposure

Imaging
area

Slower transfer to
one-dimensional
register and A/D
converter

Memory
area

A/D
converter etc.

Figure 8.2a. The
frame transfer CCD:
image down-loading.

objects. Obviously, this is desirable in video cameras, which is why amateur CCD cameras with interline CCDs contain standard mass-produced (and inexpensive!) video camera CCDs. Unfortunately, the presence of memory zones on the CCD means that less light is being captured by the chip. However, things are not as bad as they might be. CCD manufacturers, like Sony, now place tiny lenses over each pixel to capture some of the light that might otherwise have been wasted.

Camera Design

I've briefly described the operating principles of CCD chips. However, the chip is only part of the story. A CCD chip needs to be housed in a suitable enclosure and interfaced to a PC or frame store. Although CCD cameras are far more sensitive to light than photographic film, careful design and usage can

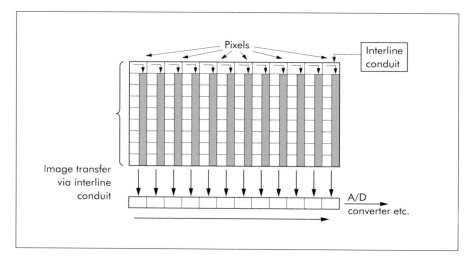

Figure 8.2b. The interline CCD: image down-loading.

increase their already impressive characteristics to the point where they become truly awesome. In the dark skies of New Mexico, Warren Offut has imaged some of the Kuiper belt objects at magnitudes as faint as 22 or 23, using a CCD camera on a 60 cm Cassegrain.

The bane of all astrophotographers and CCD imagers is a poor "signal-to-noise ratio". In astronomy, the signal comes from the stars and galaxies; the noise comes from light-pollution, grainy film and "noisy" electronic circuits. Electronic noise in CCD chips is generally referred to as "dark current" or "thermal noise". The effect of the noise builds up with the duration of the exposure and thus acts in a similar way to light-pollution as far as the final image is concerned.

Fortunately, thermal noise can be reduced significantly by cooling the CCD chip. Most amateur CCD cameras are designed to cool the chip by about 30°C. Thermal noise halves for every 7°C drop in temperature, so a 30°C drop reduces noise by a factor of around 20×.

Cooling a CCD chip is usually achieved by using a *Peltier* cooler which is a device that acts as a heat pump when electrical current is passed through it. One end of the Peltier cooler becomes cold as heat is transferred to the opposite end of the device. To dissipate the heat from the "hot" end of the device, a large metal finned heat-sink is usually attached to it (see Figure 8.3, *opposite*).

Unfortunately there is a downside to cooling the CCD chip by this amount. As anyone without double glazed windows will be well aware, when relatively

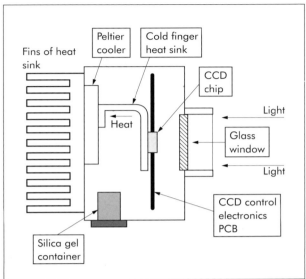

Figure 8.3.
Cross-section of a
simple CCD camera.

warm air hits a colder surface, the moisture in the air condenses out. If the temperature of the warmer air is below freezing, frost may even form on the cold surface. Obviously this effect is not helpful in a CCD camera! The solution is to seal the CCD camera body with a glass window to prevent warm air landing on the CCD chip. By making the body of the camera reasonably air-tight and drying the air inside the camera, problems with moisture can be eliminated. Drying the air is relatively easy; a substance that absorbs moisture (a desiccant) can be stored in a porous container in the camera body such that air circulating in the camera passes through the desiccant and is dried.

Periodically, the desiccant (typically silica gel granules) is removed, baked and then replaced in the camera. I recommend that the CCD camera user obtains some "self-indicating" silica gel; this kind is blue when dry but turns pink when wet and in need of replenishment. The glass window at the CCD camera not only prevents condensation and frost forming on the CCD chip, but it also keeps dust off the chip surface. However (and as you will see later) dust on the glass window can cause "doughnut" features under certain conditions.

So – what can a CCD camera do for us, and how do CCD cameras compare with photographic film and with the human eye?

CCDs Compared with Film

In the photography section, I said that gas-hypered Kodak 2415 Technical Pan film was generally regarded as the film of choice for amateur astronomers. Its fine grain and sensitivity to light are remarkable compared with faster ISO, but unhypered, conventional films. Assuming this really is the best film for astrophotography, how much better are CCD cameras?

Well, my own tests show that the interline Starlight Xpress SX (see Figure 8.4) and SXF cameras I own allow me to image stars 2.5 magnitudes deeper than I can reach with hypered Kodak 2415 and the same exposure time. This translates into a (roughly) ten-fold improvement in limiting magnitude or time taken to achieve a given magnitude with the same telescope. Even the relatively large 15 micron pixels in these devices allow me to achieve better resolution than with hypered Kodak 2415. This seems strange at first, since if you look at the Kodak 2415 data sheet a resolution of up to 400 lines/mm is claimed: by comparison, 15 microns corresponds to a $1/67^{th}$ of a mm.

However, the ability of a film to resolve high-contrast black-and-white lines says little about its ability to resolve subtle features in planetary cloud belts or on galaxies. Again, as far as my own tests are concerned, the interline Starlight Xpress SX and SXF cameras deliver similar results, in terms of resolution, as hypered

Figure 8.4. The author's Starlight Xpress SX CCD head and keyboard for comparison.

Kodak 2415 at twice the focal length; this applies to Deep Sky objects and the planets.

Interpreting these results in another way, hypered Kodak 2415 behaves like a CCD chip with a tenth of a CCDs sensitivity and with 30 micron pixels. Or another comparison: using a CCD camera, rather than film, is a bit like switching to a telescope with three times the aperture and twice the focal length – e.g. a 30 cm f/5 telescope becomes a 90 cm f/3 system!

My recent experiments with a frame transfer CCD indicate that the sensitivity is even more impressive than for an interline device, with more than a 3-magnitude advantage over hypered 2415. It is not hard to see why CCDs are becoming so popular with former astro-photographers. During the peak of Comet Hale–Bopp I took numerous CCD exposures in an attempt to determine which exposure time gave the result that best approached what I could see with the eye, looking through the same instrument on the same night. After many nights of experimenting I decided that an exposure time of 2 seconds (plus a lot of image processing) gave the nearest result (see Figure 8.5). Less than 2 seconds and the tail disappeared; more than two and the "waves" in the comets head became overexposed. The biggest problem was coping with the large difference between the faintest and brightest parts of the comet.

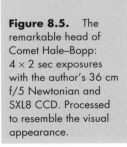

Figure 8.5. The remarkable head of Comet Hale–Bopp: 4 × 2 sec exposures with the author's 36 cm f/5 Newtonian and SXL8 CCD. Processed to resemble the visual appearance.

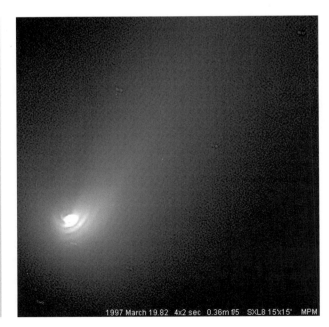

1997 March 19.82   4x2 sec   0.36m f/5   SXL8 15'x15'   MPM

The range of brightnesses that a system can tolerate is referred to as the *dynamic range*. Photographic film is at a distinct disadvantage here, especially if the astrophotographer is working at a badly light-polluted site. Before the advent of hypered Kodak 2415, astro-photographers used relatively grainy films such as Tri-X or T-Max 400. From dark observing sites these films were excellent, but they had a poor dynamic range. This meant that when light-pollution was high the film rapidly became badly fogged until the negatives became to dark too print. In addition, fogged film produces enlarged film grains and tiny star images can become lost in the grain.

CCDs have a distinct advantage over film in light-polluted areas. To start with, they have a wide dynamic range and a deep well capacity which means that even when the CCD pixels are almost saturated with light (when the bucket is full), stars which are less than 1% as bright as the background sky can easily be recorded. In addition, because the digital images from CCD cameras can be processed, it is easy to subtract all of the light pollution from the image and enhance the remainder. Because light-pollution is fairly evenly distributed over the field of view, the astronomical objects (stars, galaxies etc.) can be imagined as sitting on top of a sea of light-pollution (see Figure 8.6).

Subtracting the "sea" by image processing will leave the astronomical objects alone. Of course, if the bright

Figure 8.6.
A typical CCD histogram. The night sky sits on top of a wall (left-hand side) of light-pollution and thermal noise.

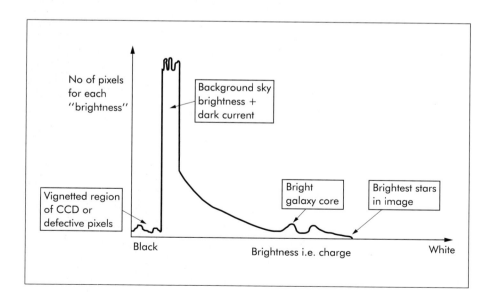

No of pixels for each "brightness"

Background sky brightness + dark current

Vignetted region of CCD or defective pixels

Bright galaxy core

Brightest stars in image

Black

Brightness i.e. charge

White

sky background is very uneven the subtraction process is complicated. Another advantage that a digital image has is that it can be added to other digital images. Thus, when a CCD is close to saturation, the exposure can be terminated and a second, third or fourth image started; all the images can subsequently be co-added with ease, revealing more and more faint stars with each addition.

Disadvantages of CCDs

The main technical disadvantage of CCDs, compared with photographic film's is their tiny size. A single frame of 35 mm film covers 36×24 mm; a typical CCD chip covers 6×4 mm. The tiny size of CCDs can make finding a specific field like looking for a needle in a haystack, especially when using a telescope without accurate positional encoders or setting circles!

I strongly recommend purchasing a flip-mirror device (Figure 8.7) to alleviate at least some of the difficulty encountered when using CCDs with long-focal-length instruments. Using such a device, the light can be directed to the eyepiece or the CCD by flipping a mirror. The view through the eyepiece is a mirror image, but much wider than the CCD field.

Figure 8.7. The author's CCD flip-mirror device + "SX" CCD camera.

On the Newtonian used with the flip-mirror device shown, the drawtube/flip-mirror/CCD system is arranged so that the CCD is always very close to focus when the assembly is inserted into the telescope drawtube (to avoid the further hassle of re-focusing each time that the system is replaced).

An associated disadvantage of CCDs is the relatively small number of pixels across the length and width of the device. Amateur CCDs typically contain over 100,000 pixels; some contain half a million pixels or more. These numbers sound impressive but they pale into insignificance compared with the amount of information that photographic film contains.

The other major disadvantage of CCD cameras for amateurs is, of course, their cost.

Image Scale and Pixel Size

The Earth's atmosphere is a seething, turbulent mass, through which light from the stars has to pass. Not surprisingly, at high magnifications this turbulence can be seen, especially when observing the Moon and planets. From a typical amateur-observing site, star images are swollen into blobs around 3″–4″ in diameter. For this reason, seeing planetary details at the diffraction limit of a decent sized telescope (0.4″ for a 30 cm telescope) is very rare and demands exceptional atmospheric stability, or a very patient observer. If the atmosphere will not allow you to image arc-second details over extended exposure times then there is little point in having a long-focal-length Deep Sky system, especially when this leads to a tiny field of view. If your CCD images show even faint stars as large and bloated, then it is time to reduce your image scale and benefit from the wider field of view. Calculating the optimum image scale for a CCD camera is simple, as pixels rather then amorphous grains are involved.

The Nyquist sampling theorem says that if you want to capture data at a certain resolution, you should sample at a resolution which is twice as fine. Put another way, if you want to resolve 4″ detail in an astronomical image, you should arrange for your CCD pixels to span 2″. Obviously this can only be achieved by optically compressing or extending the focal length.

Let's go for an example. If we want to find how may arc-seconds per pixel a system delivers, we can use the following formula:

arc-seconds per pixel = pixel size in microns/
(1000 × focal length in mm × tan (1/3600))

which reduces to:

arc-seconds per pixel = pixel size in microns/
$(4.85 \times 10^{-3} \times$ focal length in mm)

(A micron is a thousandth of a millimetre.)

For a worked example, let us take my own 0.45 m f/4.5 Newtonian (focal length 2205 mm) and a Starlight Xpress CCD with 15 micron pixels. Using the above formula gives:

15 microns/$(4.85 \times 10^{-3} \times 2205$ mm$) = 1.4''$/pixel

By comparing images I have taken with various instruments and images that Denis Buczynski has taken with a 33 cm f/3.5 Newtonian it would appear, for UK observers at least, that there is little point in using image scales finer than about 2''–3''/mm for Deep Sky images.

Denis's short-focus Newtonian, when used with 15 micron pixels, delivers $15/(4.85 \times 10^{-3} \times 1155$ mm$) = 2.7''$/mm, roughly twice as coarse as the 0.49m f/4.5 Newtonian image scale. Yet the images taken with this compact system show no finer detail when compared with those taken with the longer-focal length instrument. These results agree with the findings of other amateur CCD users at average sites world-wide. Of course professional astronomers, with telescopes at the best locations, enjoy far better conditions and can afford to use finer resolution image scales.

When choosing an astronomical telescope or CCD it is important to consider what image scale you want to work at, so that you match your system to your conditions, resolve to the limit of your site and enjoy a good field of view. An image scale that is too large has another disadvantage: if faint star images cover dozens of pixels, longer exposures will be needed to reach faint magnitudes. If we choose an image scale of 3''/mm and use a large CCD with, say, 600 × 600 pixels, we can enjoy a field of view of 3 × 600 = 1800'' or 30', or half a degree – a nice field for a CCD camera.

For wide-field CCD imaging, my 530 mm focal length Takahashi E-160 astrograph when used with a 7.7 mm square CCD with 512 × 512 pixels delivers a field of 50' square. In this case, each 15 micron pixel covers about 6''.

The examples given above are for CCD cameras with 15 micron pixels, typical for the Starlight Xpress range of CCD cameras. However, cameras manufactured by SBIG (Santa Barbara Instruments Group) have 9 micron pixels, so a correspondingly smaller focal length will be needed to achieve the same image scale. Similarly, CCDs with large pixels will justify going to a longer focal length. Of course, if you already own a telescope and a CCD camera you will have no alternative but to employ optics to adjust the focal length. Because CCD pixels are so small it is invariably necessary to reduce the focal length of all but the smallest telescopes to give a scale of 2″/pixel at the focal plane. Reworking the earlier formula and taking a scale of 2.06″/mm gives us a very convenient rule-of-thumb, namely: optimum focal length in millimetres = pixel size in microns × 100. Thus a 9 micron pixel CCD requires a 900 mm focal length. Similarly, a 15 micron pixel CCD requires a 1500 mm focal length. This, of course, is only valid for Deep Sky imaging where the average star image is, maybe, 4″ in diameter. Of course, if your tracking is very poor, you may get much larger, trailed, star images.

It can at once be seen that the required focal length, especially for the 9 micron case, is much shorter than the focal length of most medium-sized amateur telescopes. In particular, medium-sized Schmidt–Cassegrains, typically f/10 or f/6.3, work at focal lengths of up to 4 metres (in the extreme case of the 40 cm f/10 Meade LX200). As often happens in these cases, the customer demand for shorter focal lengths prompted one company, namely Optec, to develop an f/3.3 Telecompressor for f/10 Schmidt–Cassegrains, and their device has fast become a standard accessory for Schmidt–Cassegrain owners. Producing a device that can compress a focal ratio by a factor of 3, without excessive aberration, is a formidable challenge, but fortunately the small size of the CCD chip is a distinct advantage here. It is only necessary to provide a distortion free field slightly larger than the average amateur astronomer's CCD, not over the width of a 35 mm film. The Optec 0.33× Telecompressor reduces the focal length of a 25 cm f/10 Schmidt–Cassegrain to a mere 825 mm, delivering 2.2″/pixel with 9 micron pixels, i.e. just what is required. Figure 8.8 (*opposite*) shows the large spiral galaxy M81 taken with my 30 cm Meade LX200 at f/3.3. With the larger instrument, the image scale is around 1.8″/pixel with a 16′ × 24′ field. In a mere 2-minute exposure, the image reaches to well below magnitude 18 – an equivalent exposure on hypered 2415 would require about 30 minutes!

M81 0.3m f/3.3 + ST7. 2 min. 1998 Jan 31 2026-2028 UT. 24' x 16' ↑ S

Figure 8.8. M81 imaged by the author with a 30 cm Meade at f/3.3, giving a near-optimum image scale of approx. 1.8″/pixel and a field of view of 24′ × 16′.

Used with the smaller 20 cm f/10 Schmidt–Cassegrain, the standard 0.63× telecompressor will deliver 1.5″/mm with 9 micron pixels, an alternative solution for those with a better observing site than most.

So far, we have been discussing the optimum image scale for imaging Deep Sky objects, but what about the Moon and planets? Where these objects are being imaged, the goal is to resolve the finest detail possible, that is to record detail at the theoretical "diffraction limit" of the telescope. This limit can be calculated from the formula: resolution (in arc-seconds) = 116/aperture in millimetres.

As an example, a telescope of 250 mm aperture will have a "diffraction-limited" resolution of 0.46″. To put this into perspective, the planet Jupiter is roughly 50″ in diameter when at its closest to Earth and the Cassini division in Saturn rings appears roughly 0.7″ in diameter on the best amateur images (see Figure 8.9, *overleaf*).

It is not hard to see that under superb seeing conditions, a wealth of detail should be observable on the planets, even with modest equipment. The problem is that optimum conditions rarely materialise and all too often the planets look as if they are lying at the bottom of a pool of water, all shimmering and blurred. Nevertheless, the aim is to capture the detail at the limit of the telescope's resolution, however many nights and exposures this takes: all one is waiting for is

1997 Oct 18 21:37 UT 0.36m Cass f/40 2s SXL8 M.Mobberley

Figure 8.9. Saturn imaged by the author using a 0.36 m Cassegrain at f/50. The Cassini division is roughly 0.7″ wide (0.17″/pixel).

that brief moment when the atmosphere is still and unmoving for a second or so. Therefore, the image scale for planetary imaging should be arranged such that, by the Nyquist theorem, two pixels span the smallest resolvable angle at the focal plane. For a 250 mm telescope with a resolution of 0.46″, each pixel should span 0.23″ to guarantee resolution to the telescope limit. As with Deep Sky imaging, a focal length that is too long is a waste; it will result in a longer exposure time, but no gain in resolution.

So what sort of focal lengths do we need for planetary imaging with our 9 and 15 micron pixel CCDs? Unlike the Deep Sky case, here we need to consider the aperture of the telescope in question. The larger the telescope, the finer the resolution and a longer focal length will be needed to exploit the smaller angular resolution. From the formulae already discussed an extra formula can be derived, namely:

Focal length required to resolve to the telescope
limit = Pixel size (microns)/(1000 × tan θ)

where θ = half the theoretical resolution of the telescope mirror; the angle θ is given by the reciprocal of the product of 62 × the aperture in millimetres.

A worked example is always instructive (if not fun), so let us take the case of my personal favourite instrument for planetary imaging: my 360 mm aperture Cassegrain used with a Starlight Xpress SXL8 CCD camera with 15 micron pixels. In this case the angle θ is 1/(62 × 360) = 4.48 × 10^{-5} degrees. This corresponds to 0.16 arc-seconds. The focal length required is given by:

15 microns/(1000 × tan 4.48 × 10^{-5}) = 19184 mm
or 19.2 metres

In fact, for telescopes of larger aperture than this, it is highly unusual for their maximum resolution to be achieved. Thus, for large apertures, it would not be unreasonable to insert a smaller aperture figure in the above formula. The f/ratio required for the above case is obviously 19 184 mm/360 mm = f/53.

In fact, as the required focal length is proportional to aperture and as we are dealing with small angles here, a very simple and convenient formula can be derived for calculating the optimum f-ratio for high-resolution planetary imaging. By taking the aperture dependency out of the angle θ we arrive at a formula for the optimum f-ratio of :

$$\text{pixel size}/(1000 \times \tan(1/62)) = 3.55 \times \text{pixel size}$$

For an amateur telescope of *any aperture,* the f-ratio required to resolve to the telescope limit is 3.55 × pixel size. A 15 micron system, as shown above, will need to work at f/53; a 9 micron system will need to work at f/32.

Fortunately, it is far easier to extend a telescope's focal length than to compress it. Compressing leads to severe aberrations; extending does not. If you are using a Schmidt–Cassegrain, a couple of 2× or 2.5× Barlow lenses will give the required image scales. With my own 0.36 m f/25 Cassegrain, a single 2× Barlow lens produces the required f/50.

For Newtonian reflectors the situation is less happy, but eyepiece projection, or a combination of Barlow lenses and eyepiece projection, can be used to extend the f-ratios by a factor of about 10×.

Buying a CCD Camera

In the UK, the choice of which camera to buy is largely determined by whether you want a Starlight Xpress system or not! This is the dominant player in the UK. As I already mentioned, the British amateur astronomer Terry Platt developed the Starlight Xpress range of Interline CCDs in the early 1990s, after building a system of his own for imaging the planets.

Starlight Xpress

The fact that the Starlight Xpress SX system is, in the UK, a third of the price of its nearest imported rival is a major selling point.

Terry was so renowned in the UK for his planetary images that as soon as his cameras became available they were swallowed up by amateurs. In the early 1990s the alternative camera to the Starlight Xpres, the SBIG ST4, was tiny in comparison to Terry's Interline 500 × 256 pixel system. Larger cameras were twice the price, being imported from the USA. My first Starlight Xpress CCD system was purchased in June 1992 and was only the third unit sold by the company.

This original system, still available in enhanced form, stores the image in a frame store (RAM) housed in a metal box. A TV monitor connected to the box displays the image. The system was designed before PCs were even a tenth as fast as they are today: one CCD image would fill a PC's RAM, and an evening of imaging would fill a hard disk!

A system which was independent of a PC and could download an image in a couple of seconds was ideal, especially for planetary observers who might wish to rapidly capture a hundred images in one session, but only keep the best one in the framestore. With this system, it was not even necessary for the owner to lug a bulky PC outdoors, unless he wanted to store the images. A system at a public observatory, or one used for casual Supernova patrolling, could therefore be used stand-alone. This framestore system (Figure 8.10, *opposite*) interfaced to a PC via the PC's internal expansion card slots and a ribbon cable, i.e. the PC casing had to be removed to facilitate the interface.

A lot has changed since 1992; in particular, PCs are considerably more powerful and their parallel ports are much quicker. Terry's most popular cameras (the SX series) now interface to a PC via the parallel (printer) port. Image download speeds from the SX head are now well under 10 seconds (even with 486 PCs) but the framestore version is still quicker.

Another attractive feature of the SX heads is their light weight. My ST7 CCD is too heavy to easily hang from my Takahashi E-160 photographic adapter, but the SX camera causes no problems because it weighs no more than a medium-format camera.

The basic SX system utilises an Interline Sony CCD with rectangular pixels roughly 13 × 16 microns across. Although the chip has 500 × 290 pixels, Terry's original system only used 255 of the 290 (for historical reasons concerned with memory addressing). The active area of the Sony chip is roughly 6.35 mm × 4.52 mm. The optimum Deep Sky imaging focal length with this CCD

Figure 8.10. The original Starlight Xpress framestore system consisting of a PC, framestore electronics and separate monitor. (The CCD head is not shown.) Here, the system is shown mounted on the author's observing trolley.

works out at around 1400–1500 mm for approximately 2″/mm, close to the focal length of many amateur Newtonians.

Rectangular pixels mean that the image can look strange when displayed on a PC screen. Image processing packages will display each pixel as a square pixel on the PC screen (as opposed to a TV monitor). In a system with rectangular pixels, the aspect ratio (the ratio of the width to the height) of the image will be wrong and must be corrected. To prevent anomalies arising during this process (e.g. bright stars looking non-circular, with bits detached) the image should be "re-sampled" accurately by the software. A badly re-sampled image (perhaps resulting from the use of inferior software) can look distinctly second-rate.

Starlight Xpress also market a much larger chip in the SXL8 system. Although the chip in this system is only half as sensitive as the SX chip it does have square pixels, an impressive dynamic range, a good response in the blue end of the spectrum and almost photographic-quality CCD images. The SXL8 chip pixels are 15 microns square and there are 512×512 of them, yielding 7.7×7.7 mm of active imaging area.

Colour Imaging

In addition to their monochrome CCD cameras, Starlight Xpress are unique in offering a "single-shot" colour CCD camera for amateur astronomers. This camera is marketed as a colour version of the monochrome SX system and is very competitively priced.

Unlike tri-colour CCD imaging, the Starlight Xpress CCD system avoids the use of filter wheels, multiple exposures and image registration. This system exploits the fact that the chrominance (colour) resolution does not need to be as high as the luminance (brightness) resolution as far as the human eye is concerned: the same trick is used in standard colour TV transmissions, because it means that relatively little extra data is necessary for a colour image. By separating luminance and chrominance data, powerful image processing routines can be applied without distorting the colour balance.

Thus, with a relatively minor loss of sensitivity and resolution when compared with its monochrome cousin, the Starlight Xpress Colour SX head adds a new dimension to astro-imaging. The hard part is achieving perfect colour balance, and knowing when you have got it right. For example, when taking colour images from low-pressure sodium light-polluted sites, knowing just how much of the orange to subtract (even using the "light-pollution filter" software option) can be quite tricky.

The single-shot astronomical CCD is much simpler than traditional tri-colour imaging through red, green and blue filters. So much so that the reaction of most amateur astronomers is, "What's the catch?" Another typical reaction is, "Hold on, colour CCDs are used in every camcorder, so why are all the astronomical CCDs except this one using black-and-white chips?"

To deal with the second point: Starlight Xpress use interline camcorder CCDs rather than the frame-transfer devices favoured by other astronomical CCD manufacturers. These interline CCDs are readily available with a built-in "secondary colour matrix filter pattern", a "checkerboard" of yellow, magenta, cyan and green filters on the CCD chip. Frame transfer CCDs are not available with this colour filter matrix.

Taking long CCD exposures with a colour CCD is not simply a case of extending the exposure. In a camcorder the image downloading and colour decoding are performed by digital video hardware at the video frame rate (25 frames/second for the European PAL system, 30 frames/second for NTSC used in the USA). If a camcorder CCD is used for long astronomical exposures, the downloading and colour decoding are under software control, which means a not-inconsiderable investment in software development.

Although the Starlight Xpress interline CCDs are less sensitive than their frame-transfer counterparts – and the colour chip even less so (50% as sensitive as the black-and-white version) – the single-shot colour CCD takes a colour exposure in *less* time than a tri-colour frame-transfer system. This is because a tri-colour image through red, green and blue filters requires three separate exposures of long duration (especially for the blue image, as frame-transfer CCDs are relatively insensitive at the blue end of the spectrum). In addition, the strong infra-red response of CCDs means that the "red leak" of a single blue filter can result in infra-red radiation registering as blue, thus distorting the colour balance. Because of this, an infra-red blocking filter is usually required, extending exposure times even further. So, in answer to the earlier question, "What's the catch?", there doesn't seem to be one!

A further advantage of a single-shot colour CCD system is that colour exposures of fast-rotating planets (Jupiter is the worst offender) are far less stressful! At opposition, features on Jupiter's central meridian are moving at 15 arc-seconds per hour when viewed from Earth. As amateurs strive to achieve sub-arc-second detail on the planets, it can easily be seen that all three tri-colour images must be obtained within a period of a few minutes if unacceptable registration problems are to be avoided. In the UK, it is not unusual to wait an hour or more for that one second when the seeing is perfect, making tri-colour planetary imaging virtually impossible.

Recently, Starlight Xpress has added another novel camera to its range: the MX-5 CCD. This camera features a small CCD chip with an imaging region of 4.9×3.6 mm and 500×290 pixels. The most remarkable feature of the MX5 camera is its size: no bigger than a 2″ (50.8 mm) eyepiece.

Although it wasn't my intention in writing this book to compare different manufacturers' products in a value-for-money fashion, I believe that in the UK the Starlight Xpress CCD range is too competitively priced to ignore. Even in the USA it is outstanding value. *But,* if you want more pixels and sophisticated features such as autoguiding and automatic dark frame exposures, you need to look at the competitors' cameras.

SBIG and Meade

Outside the UK, the Santa Barbara Instrument Group (SBIG) and Meade dominate the amateur CCD market. Both companies' flagship cameras use the same imaging chips, the Kodak KAF-0400 and KAF-1600. The KAF-0400 is used in SBIG's ST7 and Meade's Pictor 416 cameras; the KAF-1600 is used in SBIG's ST8 and Meade's Pictor 1616 camera.

For most amateurs the KAF-0400 CCD currently represents the "dream" CCD imager; 768 × 512 pixels (9 micron) giving an active area of 6.9 × 4.6 mm. But even this chip is dwarfed by the KAF-1600 which is four times bigger, with 1536 × 1024 pixels and a 13.8 × 9.2 mm imaging area. Unfortunately price tags of $2500 and $7000 respectively mean that only a few amateurs – especially in Europe – will opt for the latter. Figure 8.11 (*opposite*) shows the relative sizes of some popular CCD imaging chips.

Although a large CCD chip is a tempting proposition, it would be wasted if some thought were not given to the optimum focal length, as discussed previously. For planetary imaging, a 1536 × 1024 pixel CCD is quite unnecessary. Even under exceptional seeing, an image scale of 0.2″/pixel is more than adequate and with Jupiter spanning 50″ at most, only 250 × 250 pixels are needed to capture the largest planet at the best possible resolution.

For Deep Sky imaging it might, at first glance, seem like a good idea to purchase a KAF-1600 system and use it at the focal plane of a long-focus instrument, say a 30 cm f/10 Meade LX200. With such a system, a field of 10′ × 15′ results, with an image scale of 0.6″/pixel. But at this scale, even faint stars will span many pixels and the limiting magnitude will suffer.

However, the KAF-400 CCD (at a third of the price) with a 0.33× Optec Telecompressor yields a field of

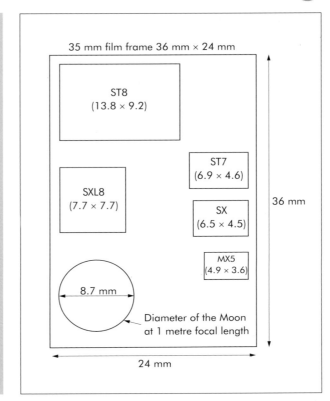

35 mm film frame 36 mm × 24 mm

ST8
(13.8 × 9.2)

ST7
(6.9 × 4.6)

SXL8
(7.7 × 7.7)

SX
(6.5 × 4.5)

MX5
(4.9 × 3.6)

36 mm

8.7 mm

Diameter of the Moon
at 1 metre focal length

24 mm

Figure 8.11. The relative sizes of CCD chips with respect to 35 mm film.

view of 15′ × 23′ and a much more efficient 1.8″/pixel. For the modest cost of a Telecompressor, a wider field and a more sensitive system results, and all for less than 40% of the cost of the very large KAF-1600 CCD! (Incidentally, the KAF 1600 cannot be used to full advantage with the Optec 0.33× Telecompressor because the chip is too big to fit within the aberration-free field.)

Auto-Slewing with a Schmidt–Cassegrain and a CCD Camera

Returning to the more affordable world of the KAF-400 CCD chip, this chip has now become the CCD of choice for amateurs world-wide. At a scale of 2″/pixel the CCD covers a very useful 26′ × 17′, a crucially important factor when one is using an auto-slewing Schmidt–Cassegrain.

Although these instruments are advertised as being able to slew to a precision of one arc-minute (i.e. putting the target right in the middle of the CCD chip), I find this very difficult to achieve in the field. In practice a slewing accuracy of 5′–10′ is more realistic, so a CCD with a large area is useful.

SBIG Autoguiding

An additional advantage of the SBIG ST7 CCD camera (see Figure 8.12) in this respect is that it incorporates a separate smaller CCD for autoguiding. This feature sets the ST7 in a different league to its competitors.

The second CCD is positioned just above the main chip and in the same focal plane. Thus, the star field seen by the guiding chip is only just outside the field of the main CCD. This is optimum because the guide stars are not badly distorted (as they might be when an off-axis guider is used), even with a Telecompressor. And there is no problem with flexure between a guide telescope and the main telescope tube. The only disadvantage I have experienced with this system is a small amount of vignetting at f/3.3, caused by the dividing

Figure 8.12. The SBIG ST7 CCD and Optec reducer.

wall between the main and guiding CCDs. However, this is easily corrected during image processing.

Autoguiding a telescope by using a CCD is not a new idea; indeed, SBIG's ST4, introduced in 1989, was intended as both an imager and a guider. The concept of controlling the drive motors of a Schmidt–Cassegrain to compensate for the drifting of a star in the CCD field was new for amateur astronomers in 1989, and initially SBIG's tiny ST4 CCD was used mostly to guide telescopes for conventional ("wet") astrophotography. An off-axis guider body was inserted between the SCT backplate and the camera, and the autoguider was then inserted into the off-axis guider, where the eyepiece would normally go.

Control software then monitored the motion of the bright guide star as it drifted across the CCD, instructing the telescope drive motors to compensate for the tracking errors and keep the star centred. The guide star does not have to move by even one pixel to cause the system to make a correction; even a movement of a fraction of a pixel will alter the amount of light falling onto pixels under the guide star, and the system will respond. The power panel of all advanced Schmidt–Cassegrains now features a CCD guider socket. (Figure 8.13).

Of course, amateur astronomers then wanted to go one step better, i.e. use a CCD autoguider to guide while a separate CCD camera took the exposure. Many amateurs eventually upgraded to a larger CCD than the ST4, but retained their original CCD camera as an

Figure 8.13. LX200 power panel. The CCD guider socket is second from the right.

off-axis guiding system. Meade's top two "Pictor" CCD cameras (i.e. the 416 and the 1616) currently include a "free" Meade CCD Equipment system, including an off-axis guider and 201XT autoguider, to remain competitive with SBIG's ST7. The ST7 has recently been joined by an exciting new SBIG product, the AO-7 Adaptive Optics system. This uses the ST7/ST8 guiding CCD to monitor star movements for rapid shifts in motion due to the Earth's turbulent atmosphere. A lightweight tilt/tip mirror then corrects for the image shifts at speeds of up to 10 ms, i.e. far quicker than the telescope's drive or a human being can react. The image is thereby corrected at a speed which even the visual observer cannot hope to detect. A two-fold visual improvement is claimed by SBIG.

With modern technology, the worst nights are most definitely things of the past, for me at least. Today, with an autoguiding camera like the SBIG ST7, I just have to go outside, open the observatory, calibrate the telescope on one or two reference stars, turn the dew zapper on and then return to do the rest of my night's observing from indoors.

Some observers, involved in Supernova Patrol work, can image one or two hundred galaxies *per night* from the comfort of a warm room, and magnitude 18 can be reached in under two minutes with a 30 cm instrument. But in practice, a lot is down to the skill and dedication of the observer. To excel in any subject you always have to be one step ahead of your rivals, and although the new technology has brought Supernova patrolling out from the impossible to the possible category for many keen amateurs, it also means there is more competition out there. The person who goes the extra mile and spends every hour of every clear night staring at his computer screen will beat the person who doesn't to the new supernova.

Other Manufacturers

We have now had a brief look at the most popular CCD imagers from SBIG and Meade, the ST7, ST8, Pictor 416 and Pictor 1616, i.e cameras using the Kodak KAF-400 and KAF-1600 CCDs, but there are plenty of other CCD cameras to choose from. Meade also make the Pictor 208/216XT models. These instruments feature a 336 × 242 pixel system and a Texas instruments TC-255 CCD chip, with an imaging area of 3.3 × 2.4 mm. SBIG used to make an ST-5 CCD camera based around the same CCD chip.

The range of relatively inexpensive Meade imagers start at around $500 (1998 prices). In addition, Celestron have also entered the CCD imaging field with their "Pixcel 255 Advanced CCD". The Pixcel 255 also uses the TC-255 chip and the whole system has been developed in collaboration with SBIG. The Pixcel utilises 10 micron pixels and is an extremely compact camera. One important consideration in the design process was compatibility with Celestron's novel "Fastar" telescope. This telescope is, in effect, an 8-inch (200 mm) Celestron Schmidt–Cassegrain with a secondary mirror which can be removed so that a CCD can be placed at the first focus, to make a sort of CCD Schmidt system. The body of the Pixcel 255 CCD camera is only 80 mm in diameter and so the light obstruction is not excessive. The Fastar + Pixcel 255 combination gives a 400 mm focal length system with 5.2″/pixel and a 29′ × 21′ field, i.e. just big enough to image the gibbous Moon. This is an attractive combination for those who want to image wide fields at a relatively low cost. The novel feature of the Pixcel 255 is that it incorporates a tiny colour filter wheel for simple tri-colour imaging.

Although SBIG, Meade, Celestron and Starlight Xpress are major players in the CCD market, there are plenty of others. In Europe, the Hi-SIS CCD range and the excellent and affordable "Quick Mips" image processing software of Christian Buil and his colleagues have a devoted band of followers.

Using a CCD Camera

You've bought it, and now you want to try it out. The best place to first test a new CCD camera and its software is *indoors*. Things can be difficult enough in the dark, damp and cold, without learning the basics of a new system. Most CCD cameras with a T-thread can take an old second-hand Pentax-threaded lens, which can be used to test the system indoors. (As I mentioned in "Camera Interfaces" in Chapter 7, the 42 mm T and Pentax/M42 threads have different pitches, but they can be interfaced in a somewhat unsatisfactory manner if you are careful not to do them up too tightly and do not damage or jam the threads.)

Assuming the camera arrives in working condition, the first possible source of confusion may be an "out of memory" warning when an image is taken. Although modern PCs have relatively huge areas of RAM – 32 MB (megabytes) is pretty well the sensible minimum requirement – problems can still be encountered, depending on how the camera software is written and how the PC is configured. Older software sometimes tries to store images in already-occupied chunks of RAM, despite the availability of free RAM elsewhere – many amateurs use old, second-hand PCs outside the observatory, rather than let their gleaming new machines get damp and rusty. This is a good policy, but your old 486 may only have 4 or 8 MB of RAM, which may simply not be enough to let your system run, especially under Windows. Windows 95 needs 16 MB minimum, and Windows NT even more. The simplest solution to this is to buy and install more RAM in your PC (consult your PC manual and dealer for details).

The thing that occasionally causes problems is an incompatibility between your camera or image processing software and the PC's video card. This is a more common problem if the software runs under DOS. It may be prudent before investing in CCD/image processing software to check whether the software is compatible with your video card, especially if you have a brand new or relatively uncommon video card. New video cards are not expensive, however. Beware also of Windows NT and some Hewlett-Packard printer software, which both delight in preventing users from doing what they want with the parallel port!

Once you have successfully interfaced your camera to the PC and managed to run the software, it's time to take a few trial exposures. Please bear in mind that an astronomical CCD camera is highly sensitive to light and in broad daylight there may be no exposure setting that is short enough to avoid saturating the camera. Trying a few exposures indoors, at night, in a dimly lit room will provide sensible levels of illumination.

Once you've secured a few images of the inside of your dimly lit room, it is time to learn a bit more about the fundamentals of CCD imaging. The first time you take a CCD exposure of a well-known galaxy or nebula you will be amazed at how much detail is revealed in even the shortest of exposures, when compared with photographic film. You will have heard that, to excel in CCD work, you need to take things called flat-fields and dark frames, but this may seem like gilding the lily, as you may well be stunned at the quality of even your first, raw images.

But you will eventually start to compare your images with those of leading amateur astronomers. To get this kind of result, you will need to understand both the limitations of their CCD systems and the role of image processing.

Understanding and Processing the Digital Image

A raw CCD image consists of individual brightness information from tens of thousands or hundreds of thousands of pixels. The electrical charge collected in each pixel is converted to a voltage and the voltage is then converted, via an A/D (Analogue-to-Digital) converter, into a binary number. (The number is represented by a sequence of ones and zeros, the fundamental binary language of computers.)

The number of brightness levels that the CCD camera can record is dependent on the resolution of the A/D converter. If we imagine a system in which only a one or a zero is available – a one-bit system – the CCD camera would then offer two levels of "grey" – white or black. (Incidentally, the word "bit" is a contraction of "binary digit.) Obviously, this would result in an image of ridiculous contrast, like a very harsh photocopy.

In practice, even the cheapest CCD cameras offer an 8-bit system, in which each pixels' brightness is recorded using eight binary digits. With an 8-bit system, $2^8 = 256$ grey levels are available. The first bit is used to represent the lowest level of variation and the eighth bit, the largest variation. Each bit therefore represents twice as large a variation as the next bit down (which is the way binary numbers work). Thus the first bit represents one unit; the second bit represents two units; the third bit represents four units etc. – until the eighth bit represents 128 units. The largest number that can be represented in an 8-bit system is 11111111 (binary) = 255 (decimal). Thus the number of possible grey levels in an 8-bit system is 256. This number of grey levels represents the minimum sensible luminance resolution level for a CCD system. If fewer bits were used, the gradual fading in intensity at the edge of a large nebula or comet would start to look false and highly contoured.

Many amateur CCD systems now use 16-bit A/D converters. These give excellent tonal ranges because $2^{16} = 65,536$ grey levels, a vast improvement over an 8-bit image. The range of brightnesses in an image is often referred to as the *dynamic range*.

Virtually all CCD camera software packages feature a histogram chart which displays the range of brightness levels from black to white along the horizontal (x) axis, and the number of pixels at each brightness along the vertical (y) axis. Understanding this histogram is of fundamental importance when deciding what image processing steps should be taken.

Unexpectedly, you will see that the brightness level for each pixel of an amateur's long-exposure astronomical image consists largely of non-astronomical data. This is easy to deduce from the histogram.

Essentially, the galaxy, nebula, comet or faint star being imaged is sitting on top of a "wall" made up of light-pollution from the sky and dark current (thermal noise) from the CCD. If you observe from a badly light-polluted site and/or if your CCD camera's Peltier cooler is inefficient, you will have significantly less signal than noise. The worse the noise, the more chance there is of faint stars being lost in it.

The situation is exactly analogous to what happens as the Sun sets. Think of the Sun as a source of optical "noise" . As it sets and the sky gets darker, more and more stars become visible: the signal-to-noise ratio improves as the Sun sets further below the horizon.

CCDs are often advertised as being extremely quantum-efficient, when compared with film – they turn far more photons into useful signal. While this is very useful for recording faint objects, the signal-to-noise ratio is just as important. A CCD user in a back-yard in New York City or central London will not be able to record faint stars as well as an astrophotographer working from a dark, high-altitude site, and the reason is signal-to-noise.

In Deep Sky CCD imaging, the prime aim of image processing is to subtract as much as possible of the noise and process the wanted part of the signal from the sky. This will make the faintest objects visible against a dark background.

Examining the raw image histogram in Figure 8.14 you can see that there is a huge spike in the histogram in the lower third of the picture. The point at which this spike starts to drop is the point at which the dark current and light-pollution end and the faintest astronomical data, sitting just on top of this noise, begins. At the far right-hand end of the horizontal axis you can see that there are just a few very bright pixels in the picture; these are due to one or two bright stars in the field.

To enhance the faintest detail in the image, we need to cut out the noise and the brightest points and then contrast-stretch the remaining data to fill the dynamic range of the new image. The result of this operation is very similar to turning the contrast up to the maximum and the brightness down slightly, on a black-and-white TV set. Suddenly, faint stars at the limit of the original

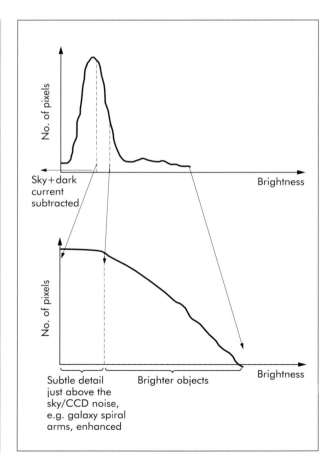

Figure 8.14.
Contrast-stretching – the raw CCD image.

image will become much more prominent, and galaxy spiral arms will become much more exciting!

A Dark Frame

The noise contributed from the dark current is subtracted by taking a so-called "dark frame", an exposure of identical duration to the main image, but with the telescope capped to prevent light hitting the CCD chip. The dark frame will consist only of the dark current plus the system readout noise.

Taking a dark frame (rather than just subtracting a brightness value across the whole image) is advisable because a CCD chip will frequently have temperature variations from one side of the chip to the other, and the dark frame includes this information. The computer software subtracts the value of the pixels in the dark frame from the values of the same pixels in the "live" frame, and the result should be an image consisting only of the wanted information.

As the user, you can then make a judgement – by viewing the histogram – about where the background sky brightness/light-pollution ends. A trial-and-error session of subtracting and contrast-stretching should then reveal the optimum points at which to subtract the pollution and stretch the remainder.

The basic contrast-stretching routine described above is one of the simplest image-enhancement routines that the amateur can carry out and is a good introduction to basic image processing. A typical example (NGC 7479) is shown in Figure 8.15 (*opposite*).

Many other image-processing routines are available in the numerous software packages available to the amateur; some of the more common routines are discussed later in this chapter.

While on the subject of dark frames, I should mention that one occasionally comes across the term "bias frame'. This is another indicator of system noise, but for an exposure of zero seconds duration: that is, *a bias frame + dark current noise for the exposure = the dark frame noise for the exposure.* Most manufacturers' software packages simply refer to the dark frame as, in practice, separating the bias and dark current noise is rarely important.

Hopefully, I've been able to give you at least an outline of the problems posed by noise, which is the fundamental limiting factor in Deep Sky imaging.

Figure 8.15.
Simply subtracting the dark frame and performing a contrast-stretch greatly enhances this image of the galaxy NGC 7479. (Image by M. Mobberley.)

Background Brightness

A term used to describe the background night sky brightness is *magnitude/square arc-second*. For example, from a very dark site of magnitude 21/square arc-second, the brightness of the night sky is equivalent to having a star of magnitude 21 in each square arc-second of the sky. In practice, stars do not have disks as small as a square arc-second (except at the finest observing sites) but this is only relevant when trying to compute your sky brightness by comparison with reference stars.

In practice, amateur CCD cameras can image objects which are less than 0.5% as bright as the background sky brightness – about 6 magnitudes/square arc-second fainter.

As a very crude guide, from a typical amateur's site with a sky background brightness of magnitude 17/square arc-second, objects as faint as mag. 23/square arc-second can be imaged. This translates to stars of around mag. 20 in practice. Of course, this is the *limiting* magnitude and will only be achieved with the CCD pixels almost saturated with light-pollution at the end of an exposure of many minutes. Shorter exposures will capture faint stars at a rate 10× faster (or more) than hypersensitised film, but the *limiting* magnitude is the absolute limit, to all intents and purposes. When the pixels are saturated, that is the end of the story!

Flat-Field

As well as the dark frame, there is another calibration frame which may be necessary to tweak your images nearer to perfection. The "flat-field" is not an essential exposure, but it can significantly improve the quality of your images, especially if you use a telescope which suffers from vignetting or a CCD camera which frequently suffers from dust specks on the glass faceplate. The "flat-field" can also correct for differing sensitivities between pixels on the CCD chip.

If the telescope + CCD system you use for imaging the night sky is pointed at an evenly illuminated target – that is, one with negligible variations in illumination – the resulting CCD image will be a record of *all* the deficiencies of the system. Less sensitive pixels will look dimmer; more sensitive pixels will look brighter. Doughnut-like shadows, caused by dust specks on the face plate, will be accurately recorded.

This information can be stored and then used to "divide out" the deficiencies in a CCD image of the night sky. In a badly vignetted image with a CCD covered in dust specks, the effect of applying the flat-field can be dramatic; in a well-designed dust-free system, the effects are less noticeable. The big problem with flat-fields is finding an evenly illuminated source which will produce a genuine flat-field. A poorly illuminated source will produce a poor flat-field, and that will be worse than no flat-field at all.

A clear twilight evening sky, imaged just before the main observing session, is frequently recommended as an excellent source of even illumination. However, if you live in a country with weather like we have in the UK, you can rarely tell whether you are going to be observing at the time of sunset! A clear sky can (and often does) become a cloudy sky in half an hour, and vice-versa!

Another option, used by many, is to place a translucent sheet of white plastic over the front of the telescope and illuminate it with a bright, diffuse light source (Figure 8.16a, *overleaf*). A trick I have used on a few occasions, if I have been doing Deep Sky or Comet imaging just before the full Moon rises, is to wait for the Moon to rise and use it as the light source. I simply point the telescope right at the Moon, with some diffuse white plastic over the front, to record a superb flat-field. As a guide, the flat-field exposure should be timed such that the CCD is at least 50% saturated.

Applying the flat-field to the image is simple in most CCD software packages. The software simply divides every pixel in the original image by every pixel in the flat-field. This operation will, at a stroke, correct all the vignetting and pixel-to-pixel sensitivity variations and any variations caused by out-of-focus dust specks on the CCD window. Strictly speaking, the raw image *minus* its dark frame should be divided by the flat-field *minus* its own dark frame, although twilight flat-fields can be extremely short exposures with little contribution from dark current.

It is important that the configuration of the telescope and CCD camera is unchanged between the exposing of the image and the exposing of the flat-field. Old flat-fields can create more problems than they solve, if things have changed. Ideally, even the specks of dust on the CCD faceplate should be in the same position for the flat-field.

The dust speck problem can be especially irritating when taking planetary images. This is because the convergent cone of light coming to focus on the CCD is much narrower in planetary work, so small specks of dust on the glass window in front of the CCD can more effectively block a higher proportion of the

a

b

Figure 8.16.
a Translucent flat-field screen for a Schmidt–Cass.
b Diffraction focuser for a Schmidt–Cass.

light. This leads to sharper and denser "shadows" on the image which are especially annoying when super-imposed on a bright planetary disk. My advice here is to keep the glass plate of the CCD as clean as possible when carrying out planetary imaging. Don't simply rely on the flat-field doing all the work, especially when powerful image-processing routines are applied to the image.

Diffraction Focusers

While constructing a flat-field cap for your Schmidt–Cassegrain, you may wish to build a diffraction focuser too (Figure 8.16b).

Because SCTs have no secondary mirror support structure (or spider), bright stars do not have star-like spikes around them when focused. These spikes, which appear in the star images of conventional Newtonians, can actually be extremely useful when taking rapid short exposures to determine the focus position of the SCT+CCD. However, by building a device like that shown, the diffraction spikes can be re-created during the focusing phase.

Useful Processing Routines

So far I have concentrated on the simplest image-processing routine, the basic contrast stretch – remember that this is not dissimilar to altering the contrast and brightness controls on the monitor.

There are many powerful routines incorporated within standard software packages and a brief mention of some of these is appropriate.

First, be aware that there is a danger in *over*-processing an image and revealing artefacts, i.e. deficiencies in the optical path, which are *not* related to the astronomical subject. The more powerful the routine, the greater the danger of producing features that are false – no more than exaggerated defects in the telescope or the CCD camera. You have been warned! Second, be aware that there is no substitute for a clear dark sky, a good telescope and good atmospheric seeing.

Unsharp Masking

For the planetary observer, undoubtedly one of the most important routines is the *unsharp mask* routine. As all planetary photographers know, the detail you see visually on the surface of, say, Jupiter, or even on the photographic negative, is *not* the detail you see on the final print, even under perfect seeing conditions.

The problem lies in the limited dynamic range of the photographic print. High-contrast photographic paper will enhance the detail in Jupiter's belts, but the darker-limb regions of the planet disappear and the centre of the planet is whited out! The photographer's answer to this is to prepare a so-called *unsharp mask* and to use it to reduce the range of brightnesses falling on the photographic paper. A better term than "unsharp" might be "blurred".

Imagine you have a highly de-focused image of Jupiter in which all of the fine details in the belts and zones have disappeared. Each tiny point in the image is therefore at the average brightness of all of the original nearby points. Although such an image might seem to be of little use, it is actually very useful as a measure of the large-scale brightness variations on the planet. By subtracting the majority of this blurred, or unsharp, mask from the original image, the large-scale brightness variations are removed "at a stroke". Thus, sticking with the example of Jupiter, it is possible to dim the overexposed middle of the planet and boost the invisible limb-darkened edge.

The amount of detail that this technique can reveal is quite remarkable and, in many ways, it mimics the view seen by the human eye, with its impressive dynamic range. (I always find that Jupiter looks quite unspherical to my eye; as well as the ability of the eye to cope with a huge range of brightness levels, I also feel that the contrast between the limb of the planet and the black sky, effectively brightens the darker-limb regions, adding to the "flat-planet" effect).

Although astrophotographers used the unsharp mask technique well before the computer era, it involved a considerable amount of darkroom work and film development. Nowadays, an unsharp mask can be produced and applied in a matter of seconds. An example of unsharp masking is shown in Figure 8.17 (*opposite*).

Some astro-software generates an unsharp mask and leaves the user to subtract it from the image; essentially this is simply acting like a median filter, see below. Other software, such as the Starlight Xpress "Pixwin", allows the degree of subtraction to be varied along with the "radius" (the radius, in pixels, over which the averaging is calculated), resulting in a powerful single-shot unsharp mask process.

The unsharp mask need not be restricted to planetary images. Other objects, such as the Andromeda galaxy,

Figure 8.17.
A single unsharp mask subtraction can dramatically enhance detail in Jupiter's belts and zones [M. Mobberley].

benefit in having their brightest regions subdued relative to their faintest regions. Unfortunately the unsharp mask subtraction tends to leave dark haloes around bright stars, which are especially noticeable when the stars are superimposed on a faint extended object.

Deconvolution

A more powerful technique, one which often yields a similar result to the unsharp mask process, is the so-called *maximum-entropy deconvolution* algorithm. A more meaningful title might be "maximum information de-blurring" algorithm.

Although the end result looks similar to the unsharp mask process (including haloes around bright stars), the method involved is different and iterative. Essentially, the algorithm enhances the original image a bit at a time, checking the original and processed images for statistical anomalies, correcting them and then going for a further, more ambitious, de-blurring attempt. After a dozen or so iterations, the maximum amount of information has usually been extracted.

This is a powerful technique which is becoming ever more attractive now that PCs have become so fast; only a few years ago, deconvolving an amateur CCD image might have taken 10 minutes or more; with a 200 MHz machine it now takes 1 or 2 minutes. A Maximum-Entropy Deconvolution package suitable for amateurs is marketed as *Hidden Image* by the Sehgal Corporation.

Median Filters

Amateur CCD users often encounter situations where images appear noisy on a small scale. Single pixels may appear black or white owing to electronic readout anomalies or cosmic ray impacts on the chip.

In these situations a *median filter* will often prove invaluable. Essentially, as its name suggests, the median filter performs an averaging function, based on comparing the intensities of nearby pixels. If you want to eliminate single, dead or saturated pixels, or just make the image look smoother, the median filter is highly recommended – but if over-used it can reverse the good work done by the unsharp mask or maximum-entropy processes and blur the image.

Non-Linear Contrast-Stretch

The large dynamic range of astronomical targets like comets or galaxies frequently presents serious problems for the amateur astronomer. A simple contrast-stretch certainly boosts the detail in the faintest regions, but the brightest regions often end up saturated – that is, whited out.

One solution to this problem is to experiment with a *non-linear stretch* in which fainter regions are boosted, while bright regions are suppressed. This may sound similar to the unsharp mask process, but the non-linear contrast-stretch processes each pixel without regard to the brightness of neighbouring pixels. It is ideal for images where the difference in brightness of different features is considerable.

Numerous other processing tricks and tools exist but are largely beyond the scope of this quick guide to CCD imaging. My advice is, as with everything, that practice makes perfect and, with experimentation, you will soon determine the best techniques for use with your system.

Image Formats

Once you have fully processed your image you will want to save it to your PC's hard disk. Many CCD cameras have their own unique file format, tailored to suit the camera and software provided by the manufacturer. However, the desk-top publishing (DTP) world has its own standard formats and you will need to convert the image you have processed into one of these DTP formats if you want to share it with other amateurs, e-mail it to friends/magazines, or post it on your World Wide Web (WWW) home page.

In any case, you may want to convert your image into a DTP format to carry out image-processing routines not available with the camera software (such as adding a caption or logo, or converting to a false-colour palette).

Commercial packages such as *Paintshop Pro* or *Adobe Photoshop* are widely used by amateur astronomers to refine their CCD images.

There are a variety of standard DTP image file formats in use: in practice, the most popular file formats are TIF, BMP, GIF and JPEG. A fifth format, FITS, is uncommon in the DTP world, but of great interest to astronomers.

TIF or TIFF (Tagged Image File Format) used to be the standard file format for DTP applications, but the file size is unnecessarily large, with hundreds of kilobytes needed to store even modest CCD images. BMP (Bitmap) is a superior format but, again, the file size is large.

The GIF file format is much more economical of space and was originally promoted by the e-mail service provider *Compuserve* (in the days before the Internet). Obviously a small file format is highly desirable when sending a file via a modem: a large file takes more time to send – and time is money! An astronomical GIF image file is typically one-eighth or one-tenth of the size of a TIFF or BMP file. The size reduction is achieved by compressing the data. Information is not lost, just recorded more compactly. As an example, in an astronomical image much of the sky may be black, or the same dark shade of grey. In a simple file format, each pixel is separately specified (the equivalent in English of "pixel 1 = black, pixel 2 = black, pixel 3 = black, etc".). In a more economical file format, a row of equally dark pixels can be indicated by a very simple code (in English, "next twenty pixels = black"), which saves a lot of space.

The JPEG (or JPG) format also uses image compression to reduce the file size. For monochrome astronomical CCD images I generally find that my JPEG and GIF files of the same astronomical targets have a similar size. The JPEG system uses a mathematical transform to compress the data and create an economical model of groups of pixels. Note that this file-compression technique is not "lossless", and subtle data can be lost when the image is compressed.

Most DTP packages such as *Adobe Photoshop* offer a quality factor when saving images in JPEG format. It enables you to make a choice between file size and image quality. JPEG images saved with the quality set to "excellent" might be compressed by a factor of, say, 4, when compared with an equivalent TIF image; with the quality set to "fair", a 10 or 12 times saving might be achieved (similar to GIF); and with the quality set to "poor" a compression factor of 20 or 30 times can be achieved. Many amateurs will wince at the prospect of detail being lost in the compression process, but in practice the loss of detail for a compression factor of $10\times$ is negligible. The JPEG system offers impressive compression ratios even on highly detailed full colour landscape shots.

The related MPEG compression transform allows impressive *video* compression with multi-frame videos occupying little more space than a single image.

Finally, there is the FITS image format. FITS stands for Flexible Image Transport System and flexibility is its most important feature. The standard was finalised by professional astronomers in 1981 and it was designed for storing the large amounts of scientific data contained in astronomical CCD images. This is not a format that the non-astronomer will encounter very often. The key strength of FITS is that it is regarded as *the* standard file format for astronomy (many of the best amateur CCD software packages can load or save FITS files). Moreover, it can be used to store images of *any* size, and the image file can store astronomical image data in the ASCII character format.

Despite its astronomical heritage, many amateurs never use FITS because it is not an economical format, nor does it appear in most commercial image processing DTP packages. Nevertheless, it is a scientific format and it is extremely flexible. The ability to save valuable image data within the file is very useful, especially if you do a lot of serious work (like astrometry or photometry) where preserving the original scientific data is all important.

Whatever format you choose to save your processed/final image in, it is always good practice to keep the original raw, dark and flat-field images in unmodified form. The original is the only true record of what you actually imaged.

Figure 8.18 (*overleaf*) shows a flow chart to show how I processed one of my images, from the raw image to e-mailing it off to colleagues.

Astrometry

Even as late as the 1980s, the only amateurs who were able to measure the positions of comets and minor planets were those with access to so-called "measuring engines". These "engines" were precisely machined platforms on which a photographic plate, or a negative sandwiched between two glass plates, could be measured by a travelling microscope mounted above the film (Figure 8.19, *overleaf*).

Because it was intended that they should achieve a measurement accuracy of an arc-second or so on the sky, the photographic telescopes (astrographs) used for astrometry needed to have focal lengths in excess of one metre, and the measuring engines needed to have a repeatable precision of better than 10 microns (a hundredth of a millimetre). Few amateurs had access to such precise measuring equipment or the skills needed

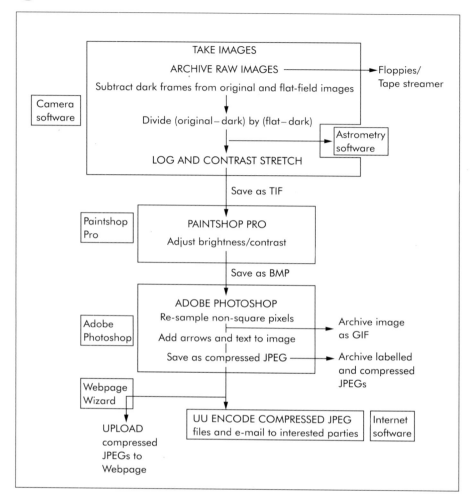

to manufacture it. In addition, astrometry was not a priority for most professional astronomers. As a result, a small number of amateurs became crucially important whenever a new comet was discovered, for they were frequently the only people likely to measure the comet's position and enable an orbit to be determined.

In the UK, during the 1950s, 1960s and 1970s the amateur astronomer Reggie Waterfield became one of these crucially important amateurs, totally dedicated to comet astrometry, despite being confined to a wheelchair for most of his life.

In the USA, comet astrometry was almost non-existent among amateurs prior to the CCD era; only Japan and Italy boasted significant numbers of astro-

Figure 8.18. The original, if convoluted, methods used by the author to take, process and e-mail images to colleagues. I now use a simpler sequence, but this does serve to illustrate how complex things can become.

Figure 8.19.
Home-made travelling microscope for astrometric measurements.

metrists. On several occasions in the 1980s, when the Japanese and Italian astrometrists were clouded out, bright (better than 10th magnitude) new comets were discovered and went unmeasured for a week or more after discovery, owing to cloud cover over the vital parts of the world.

The CCD revolution of the 1990s swept aside these problems because CCD astrometry is far easier to perform than photographic astrometry. Here's why.

First of all, the pixels on a CCD are laid out on a precise grid. The pixels themselves are typically 9 to 15 microns in size, which is about the same size as the accuracies required for photographic astrometry. In fact, positional accuracies *smaller* than the pixel size are quite achievable with CCD astrometry. (Because even a faint star is not a point source, the light spreads out over many pixels provided the image scale is good enough. The distribution of light intensity seen by the pixels can enable the software to calculate the star's centre to a *fraction* of the size of a pixel. Thus, if an image scale of 2″/pixel is employed, the position of a faint star, which is not saturated, can be determined to an accuracy of 0.5″ or better.)

Secondly, the availability of CD-ROM based star catalogues such as the *Hubble Guide Star Catalogue* mean that even narrow CCD fields have enough stars to use as astrometric references. With a PC running a modern astrometric software package (such as the *QMips* package from Christian Buil and colleagues), a few clicks of a mouse on the target and reference stars followed by comparison with the Guide Star Catalogue and, before you know it, the RA and Dec are produced and in a format ready for e-mailing to the Minor Planets Center in Boston! These days new comets are in no danger of being lost, in fact so many measurements are submitted that an orbit for a new comet is typically available within days of the discovery.

Amateur astronomers are discovering and measuring the positions of new asteroids at an alarming rate – when measuring asteroid positions is as easy as clicking a mouse button, the sky is the limit!

Some image processing packages include astrometric software. I already mentioned *QMips*, and there are plenty of other packages such as *CCD Astrometry*, *Astrometrica* and *Computer Aided Astronomy*.

Given the appropriate software and a CD-ROM star catalogue, amateur astrometry is a relatively trouble-free pursuit these days. Unfortunately, there are rather more pitfalls when it comes to CCD photometry.

Photometry

CCDs have various undesirable characteristics where photometry is concerned. Firstly, they can exhibit considerable non-linearity, especially when they are near to saturation; this is primarily due to the anti-blooming gates which drain charge away from bright stars to avoid them "bleeding" across the image. If you are solely interested in photometry, CCDs without anti-blooming gates can be acquired.

Secondly, they need to be filtered because they are sensitive to wavelengths of light from the blue end of the visible spectrum right through to the near infrared. Standard photometers, as well as the human eye, have much narrower passbands

This can cause problems: stars may show large fluctuations in brightness in a narrow band but remain relatively unchanged across the wider spectrum. An example of this was the unusual nova discovered in Cassiopeia in December 1993 by Kanatsu. This nova exhibited a *DQ Her* type fade owing to dust emission blocking the light output. Because longer-wavelength, "near infra-red" light penetrates dust far more easily than the shorter, blue wavelengths, the CCD magnitude of this nova was far brighter than the visual magnitude during the fade period.

As far as the non-linearity aspect is concerned, it is essential to determine by experiment just how linear your particular camera is (or isn't). Most CCD cameras come with software packages that offer a photometric option and even give an impressive two decimal place magnitude comparison when a star is clicked on. However, such a measurement is meaningless unless the manufacturers have taken the camera's linearity into account and a photometric filter is used. And if the star being measured is faint and only just above the noise floor, the uncertainty in the magnitude will be considerable (for example, mag. 18 ± 1 magnitude, which means the star is somewhere between mag. 17 and mag. 19!).

Professional astronomers with fully calibrated and filtered CCD systems (see the section on "Filters" in Chapter 7) can achieve photometric precision of 0.01 magnitude with a single comparison star in the field; however, even advanced amateurs struggle to achieve 0.05 magnitude precision. Note that, as a rough guide: 0.01 mag. = 1%; 0.1 mag. = 10%.

Unless you are sure of the linearity of your particular system, the safest plan is to do what most visual observers usually do, which is to use two reliable comparison stars in the field. Unfortunately, the photometric accuracies of the stars in the *Hubble Guide Star Catalogue* are surprisingly *not* completely reliable; in some cases they are one or two magnitudes in error! A star field containing photometric sequence stars should always be used where possible. Fortunately, many of the well-observed variable stars have reliable photometric sequences.

Chapter 9
Image Processing, Planetarium and Telescope Control Software

In the preceding chapter I talked about various image-processing *routines*. Of course, not all of these features are available with every software package. The purchaser of a CCD camera soon finds out that the software that comes with the camera does not necessarily come with the routines that he or she wants.

For example, I am especially interested in astrometry, simple photometry and stacking together images to allow for a comet's motion. I would also like a maximum-entropy algorithm included in the package, multiple windows to see the before and after results of processing, compatibility with all standard computer video cards, and the ability to load SBIG and Starlite Xpress file formats. A good pixel re-sampling routine for non-square pixels would also be nice, together with the ability to add text to an image and save it in a variety of file formats.

Most CCD camera software packages only offer a few of these features. The frustrated user ends up using three or four different packages before ending up with anything like the desired end result. I have used a variety of image processing software packages from Starlight Xpress, SBIG, Mira and Christian Buil as well as a package called *IMG32* written privately by Nick James, a keen UK amateur astronomer.

My personal favourite combination for processing my images is currently Christian Buil's QMips (Quick Mips) 32 system. The package is also very inexpensive. QMips 32 has almost every feature (except multiple windows) that I want, and the latest version can import both SBIG and the latest Starlite Xpress formats. It also features a superb set of contrast-stretching slide bars which adjust the contrast of an image while you watch in a reduced window at the bottom right of the screen; I find this feature extremely useful. To add labels and convert to the much more economic JPEG format I export from QMips to TIF and then use Adobe Photoshop to convert from TIF to JPEG and add labelling.

QMips also seems to be compatible with every video card I have used. This is important as many image-processing packages work under DOS and not all video cards designed for MS Windows 95 (and especially Windows NT) will allow DOS-based image-processing software to work. Before buying an image-processing package, be sure to check with the dealer whether it will run with your operating system, the amount of RAM you have and your video card.

Now let's have a look at some image-processing software in more detail.

QMips 1.81 by Christian Buil

I have already said I like this package. One of its most powerful features is the way in which different functions can be performed automatically with only one or two key strokes. Perhaps the most extraordinary example of this is the astrometric package which can identify the image star field, measure the position of the target object and even compose the correctly formatted message to the MPC (Minor Planets Center)! It will even do this for a sequence of images in which there is a moving asteroid – it identifies and measures the object that is moving. This is powerful stuff!

Most amateurs carrying out astrometry will end up using a variety of packages and spending an hour or two reducing the image and e-mailing the result to the MPC. With QMips the process can literally take only minutes.

Producing the optimum "pretty picture" from your raw, dark and flat-field images is also very slick once you have carried it out for the first time. There are over 200 commands in total; most are accessed by a command line rather than pull down menus. Ugly but fast – some people hate command lines, but I quite like them.

Other features rarely found in other software packages (and never at this price, as far as I know) are the ability to produce a cartographic projection map from CCD images of a spherical planet like Mars, blink comparator functions, maximum-entropy/Lucy Richardson "de-blurring" routines and special comet image-processing routines. QMips version 1.81 is one of the few packages to offer importing of SBIG, HiSis and Starlight Xpress formats – a reflection, perhaps, of the European origin of this software.

Disadvantages of this package? Well, my constant gripe with all image-processing software is that I would like to have multiple windows open at the same time. In addition, examining the directories from inside QMips necessitates remembering what your file name was called, unless you have re-saved it in the native QMips "PIC" format. However, all in all I believe QMips 1.81 to be one of the most powerful packages around, as well as being one of the cheapest.

MIRA AL by Axiom

MIRA AL is the only image-processing package marketed by Sky Publishing. Although all software packages require some form of learning curve, I found MIRA AL's curve steeper than most. MIRA is a professional-level image-processing and analysis software package which has been altered to suit the amateur's needs and budget.

However, a quick browse through the hefty MIRA user's guide soon reveals a minimum of astronomical dialogue, even if the processing routines are fully described. The menu-driven graphical user interface will appeal to some far more than a command line system, but most amateurs will find the lack of user-friendly astronomical advice frustrating. In the advertising blurb, MIRA AL is "used by more professional astro-imagers than all other commercial astronomical image-processing programs combined". I quite believe this claim, as MIRA's ability to handle, auto-scale and align large images and its powerful astrometric, photometric, frame arithmetic and filter routines are attractive to professional and semi-professional astronomers.

MIRA AL is in my view ideal for a small University observatory with, say, a 1 metre telescope, 1024×1024 pixel CCD chip and a photometric observing program. But QMips is a much friendlier package at only a third of the price.

Hidden Image by Sehgal

The Sehgal corporation were the first to offer maximum-entropy deconvolution (MED) image-processing (see Chapter 8) in an affordable image-processing package. To many astronomers, MED and the even more powerful Lucy Richardson routines are the ultimate tools for sharpening blurred or out-of-focus images. When hidden image was first introduced, the then-standard 66 MHz 486 PC might churn away for many minutes trying to deconvolve a 500×300 pixel CCD image. But a 200 or 300 MHz Pentium-based computer finds the task is far less tedious! Hidden image also features Fourier filtering, dark, flat and bias frame calibration, histogram equalisation and full 32-bit image-processing. It's certainly worth looking into.

CCD Astrometry

Now for some good news – some useful astro-software that is free! CCD Astrometry, by John Roger, has been around since the mid 1990s. It was the first commercial program to integrate CCD images to the *Hubble Guide Star Catalogue*, enabling accurate astrometric positions of objects to be obtained with only a few minutes' work. With image-processing packages like QMips and MIRA including astrometric reduction as part of the deal, dedicated astrometry packages like CCD Astrometry and Astrometrica are becoming financially non-viable. Hence, to download CCD Astrometry, visit the website at:

http://ourworld.compuserve.com/homepages/johnerogers

Paintshop Pro

This useful shareware package can often be found inside those free CDs which are given away with computer magazines. Paintshop Pro can be used free of charge for 30 days after it is installed; after 30 days the user should purchase the licensed version (which is modestly priced).

Most CCD cameras come supplied with software which will, as a minimum, allow dark frame subtraction, flat-fielding and simple contrast-stretching. If your prime interest is in producing "pretty pictures" following dark frame subtraction and flat-fielding, and you are not interested in photometry or astrometry, then Paintshop Pro plus the supplied camera software may well be the simplest method of achieving your goals. All CCD cameras allow images to be converted to a DTP image format; TIF is usually the one chosen. Paintshop Pro can import all of the popular formats with the exception of the astronomical FITS format.

Features available with this shareware package include:

- Contrast and brightness adjustment.
- Rotate, flip and mirror functions.
- Sharpening, blurring and user-defined filtering.
- Gamma and colour correction routines.
- Images from Paintshop Pro can also be easily printed to an inkjet or laser printer.

The only feature of Paintshop Pro I dislike is the resizing routine. Some of the interline Starlight Xpress CCD cameras feature rectangular pixels. This means that when the image is saved to an image format such as TIF, the aspect ratio is incorrect. Resizing the image should be carried out by re-sampling the pixels correctly, otherwise stars can become noticeably distorted. I have never been happy with the resizing facility of Paintshop Pro.

Adobe Photoshop

Probably the best known of the professional image-processing packages, Adobe Photoshop has been used by many amateurs to post-process their best images prior to magazine or web-page submission.

The main disadvantage of Adobe is the cost, especially when compared with shareware packages such as Paintshop pro. However, if your aim is, for example, to manipulate a scanned colour astrophoto (e.g. of a solar eclipse) for the front cover of a magazine, or to join four tri-colour CCD images into one big image, then Adobe Photoshop is for you.

Adobe can handle big colour images, allow easy manipulation of the image and add a text legend to the image. Scratches on a photo can be "air-brushed" away and the package can deal with as large an image, with as many colours/grey levels, as your graphics card can handle. Needless to say, re-sampling of pixels is performed correctly! If you have aspirations to be a leading astrophotographer with pictures on the front cover of *Sky & Telescope*, Adobe may well be the best package for you.

Printers and Scanners

No section on image-processing would be complete without a mention of printers, scanners and associated technology.

Even as recently as the early 1990s dot matrix printers were the rule and it was impossible to obtain a high-quality hardcopy of a CCD image. However, as we moved into the late 1990s, two important changes took place. First, almost everyone in astronomy acquired a PC and DTP software, and so did not strictly speaking need a hard copy.

Second, laser and inkjet printers became so good that hard copies are now (1998/99) approaching photographic quality! Advances in the performance of printers are currently so rapid that it is difficult to write anything meaningful that won't be out of date before this book is published. But at the time of writing the *Epson Stylus Colour 600* is an affordable, high-quality 1440 d.p.i. (dots per inch) inkjet printer which has found its way into many CCD camera owners' homes. The quality is approaching that of a photographic print and is more than good enough to reproduce the pixellated CCD images produced by amateur astronomers.

The Epson is not the only printer capable of near-photographic quality: check out some of the *Hewlett-Packard DeskJet* Printers too. The best plan when choosing a printer is to ask other amateur astronomers what printers they would recommend, rather than just work from the manufacturers' advertising blurb. Two printers with the same advertised resolution (d.p.i.) can produce different quality prints.

You can buy special "premium glossy" paper for use with inkjet printers, and the results obtainable can be excellent, very similar to a photographic print, but cheaper and – more important – fully under your control.

Scanning Photographs

I have already mentioned that you may wish to manipulate scanned photographs using a powerful image-processing package like Adobe Photoshop. So as well as a good inkjet printer, the amateur astronomer interested in "wet" astrophotography could also use a good scanner. Fortunately scanners are commonplace these days and capturing all the data from a print is possible, even with a relatively inexpensive scanner. Flat-bed A4 scanners will provide good scans – look at the *Epson* and *Hewlett-Packard* ranges – and even "hand-scanners" can scan a print quite well.

The problem with this technique is that it depends on the print (which may be less than perfectly done) as a starting point. Capturing the data from an original negative avoids this problem, but demands a more expensive scanner. However, this service may be performed easily by many high street chain stores and photo-shops.

A common service available these days is to provide a photo CD-ROM master, produced by scanning the negatives of a 35 mm film. A "photo CD master" disk can hold about a hundred 35 mm images. With the Kodak scanning system the images are stored using the Photo YCC colour encoding metric, which stores data at multiple levels of resolution in units called IMAGE PAC files. Most modern PCs with a CD-ROM drive can read photo CD files via software such as Adobe Photoshop.

The highest resolution available with the Kodak system is the base ×16 resolution of 2048 × 3072 pixels; each file at this resolution takes up 18 MB.

The standard "base" resolution is a more manageable 512 × 786, giving a 1.1 MB file size. At the time of writing, the standard cost of committing a 35 mm film onto a photo-CD (at the time of film processing) is roughly double the cost of developing and printing the same film.

Planetarium Software

Many amateurs will also be on the lookout for good *planetarium* packages – software that allows you to plot the stars, planets and comets as seen from your part of the world.

Many serious amateurs will also want to control their LX200s from a planetarium package and to view the pictures from the CCD camera on the PC at the same time. There are a number of features to look for:

- Will the package allow me to find everything, including the 30,000+ asteroids known?
- Can new comet discovery elements be easily imported to allow me to bring new objects into the planetarium database?
- Can I control my telescope, the CCD camera and the motorised focuser with one or two packages at the same time?
- How accurate are the positions given for asteroids and comets?
- Can I import the compressed Palomar Sky Survey (Real Sky) images into the planetarium package while it and the CCD imaging software and the telescope control software are running (e.g. for comparing the field to mag. 19 in real time)?
- Can I print star charts easily from the software?

These are important questions, and the answers will be different for each software package. There are so many available these days that space does not permit a review of them all in this book. However, good software soon attracts attention and amateurs quickly learn what to buy and what to avoid; there are a few exceptional packages which deserve a mention.

If you are looking for a visually impressive Multimedia astronomical experience where you can travel out to the planets and roam the solar system, witness its birth and enjoy over 700 stunning astronomical images, then I recommend *Redshift 2* by *Maris Multimedia*.

Redshift 2

Redshift 2 is not a full-blown planetarium package as such, but its main attraction is a dynamic solar-system model with 3D planets and moons accurately represented, i.e. you can travel to the planets and admire the view, as if you were in a spaceship. The database contains 250,000 stars, 40,000 Deep Sky objects and 5,000 asteroids as well as dozens of animations/movies with soundtracks. A fun-to-use, award-winning and inexpensive educational package, Redshift 2 works with Windows on a PC, or on a Mac.

The Sky

If you want the best planetarium package with a host of impressive features and with LX200 telescope and focuser control, then my prize goes to *The Sky Level IV* by *Software Bisque*, designed for use with Windows '95 or NT. Version 4.0 includes the Hubble Guide Star Catalogue's 19 million stars and non-stellar objects, more than 100,000 multiple and variable stars and Deep Sky objects; more than 750 colour images of the planets and Deep Sky splendours, isophotes of nebulae/extended objects, enhanced star chart printing, eclipse/occultation simulation and solar-system animation. *The Sky* can accurately depict the night sky for any date from 4713 BC to 10,000 AD for any location on Earth. *The Sky* works seamlessly with the *Real Sky* CD set produced by the Astronomical Society of the Pacific and the Space Telescope Science Institute.

Real Sky

Real Sky is a 9 CD-ROM set containing the Palomar Observatory Sky Survey. At the click of a mouse you can call up images to around magnitude 19 at a similar resolution and depth to those taken at 1–2 metre focal lengths with amateur telescopes and CCDs.

Needless to say, there is an enormous amount of data on the original Palomar Sky Survey glass plates, and in compressing this data by a factor of 100:1 some of it has inevitably been lost. Nevertheless, it is still a highly useful tool, especially for those involved in supernova searching or other patrol activities. *Real Sky* is available from the Astronomical Society of the Pacific (ASP) and, at the time of writing, a second CD-ROM set, covering the Southern Hemisphere, has been released. *The Sky* also allows you to import and overlay *Real Sky* images with relatively little fuss.

Guide

If you want the best-value planetarium package around, with basic telescope control and super-accurate sky positions, I can strongly recommend *Guide*, currently version 6.0, by *Project Pluto*. At a mere $89 (1998) it is an absolute bargain. Although the graphics are not in the same league as those featured in *The Sky*, the software is extremely user-friendly, always an attractive feature. Like QMips 32, I have yet to find a PC that *Guide* would not install and easily run on.

Upgrading to the next version of the software has always been ridiculously cheap. The only feature I can't get to work on my version of *Guide 6.0* is the *Real Sky* importing routine, although the manual does warn you that this operation is still somewhat fraught (or impossible). Visit the *Project Pluto* web site at http://www.projectpluto.com.

The features of *Guide 6.0* are:

- LX200 or Sky Commander Control.
- High precision (at least 1 arc-second for most objects).
- 18 million stars, 31,000 variables, 15,000 suspect variables, 101,000 galaxies and 1,200 open clusters.
- 30,000 asteroids.
- All the satellites of Mars, Jupiter, Saturn and Uranus.
- French, Italian, German and Spanish language options.
- A highly intuitive user interface (DOS or Windows).

Highly recommended!

Megastar

Midway in price between *The Sky* and *Guide 6.0* is *Megastar*, which contains features of both. This is a very popular package among visual Deep Sky observers, although *Guide 6.0* contains just as many objects in a less expensive package. The graphics supplied with Megastar are, to many, more appealing than those in *Guide 6.0* but do not stand comparison with those of *The Sky*. I suspect that Megastar is being squeezed between those who want a value-for-money planetarium package (i.e. *Guide 6.0*) and those who want the ultimate in graphics performance and features (*The Sky*).

Planetaria and associated software packages date quickly, but a brief mention of my top 1998 packages is worthwhile.

Starry Night Deluxe

Starry Night Deluxe was the winner of 1997's "Most Elegant" Human Interface Design Excellence award by

Apple. Although it does not feature all the huge databases and features of *Guide* or *The Sky,* it is a very user-friendly package with quality graphics and an affordable price. As with the more powerful packages, you can also control your LX200 from the planetarium menu.

NGCView 5.02

NGCView 5.02 is designed as a tool for the astronomical observer, combining an observational planner and an observing log. You type your object preferences and observing location in and the program produces a list of the most suitable objects currently observable. An observing history "filter" can also be engaged so that objects on your regular program can be prioritised – very useful.

Universe Explorer

Universe Explorer is another package in the *Redshift* mould. This is a very inexpensive CD-ROM set which incorporates a simple planetarium package as well as a 3D virtual landscape Mars Rover.

Voyager II

For the Macintosh user the long-lived *Voyager II* is still available, now at version 3.0. This package is looking rather dated now, and expensive, but is still popular among Mac users. The main features are the solar-system tour and the plotting of objects such as Halley's Comet.

Comet Explorer

Comet Explorer is a CD-ROM from Cyanogen Productions. We have already mentioned the spectacular Hyakutake time-lapse video produced by Peter Ceravolo and colleagues earlier in the book. Well, the sequence is also available on CD-ROM, as well as a host of other comet images and shareware planetarium software *Earth Centred Universe*. Worth it for the time lapse comet images alone!

The Astronomer Software

Serious observers often prefer software with the minimum of graphics, and the simplest but most accurate data. A number of such inexpensive programs have been developed for *The Astronomer* magazine (see address in Appendix 1) by Nick James. Software available works under DOS and includes:

> Eph.exe – A comet/asteroid ephemeris program with RA/Dec and motion columns
> Horizon.exe – A comet-observing altitude planner which reads the output from Eph.exe
> Chart.exe – A package which compiles charts from the *Hubble Guide Star Catalogue*
> Findvga.exe – A package which displays charts compiled using "Chart.exe"
> Ast.exe – A package which lists the asteroids at a specific RA and Dec

If you are unimpressed by flashy graphics and just want accurate ephemeris data/star charts, these very inexpensive packages may be what you require.

Planetarium Telescope Control

Many amateurs, myself included, have decided to go down the remote observing route, controlling the telescope via a planetarium package and retrieving the CCD images with CCD software running on the same PC.

A CCD camera that takes automatic dark frames is extremely useful in this respect as it means that a rotating shutter moves over the CCD rather than the observer having to leave the house and cap the telescope. Some observers think that remote observing is simply *not* astronomy: however I disagree. I've spent quite enough evenings outside frozen to the bone; on those Winter nights I can simply do far more imaging from indoors in the warm. When I feel like it I can observe visually from outside; that's my choice and no-one else's!

A Meade LX200 and, more recently, a Celestron Ultima can be controlled by the most popular planetarium packages. The data to and from the telescope is transmitted and received via the PC's serial port at around 9,600 baud (bits of data per second). PCs use the RS232 system to communicate with the serial port and this is fairly reliable out to a distance of 30 m (100 feet) or so using standard telephone cable wiring.

At longer distances – unless special driver chips or low-resistance cabling is employed – the communication will become unreliable. A block diagram of a typical remote control telescope/CCD system for a Meade LX200 is shown in Figure 9.1.

Figure 9.1. Remote control of a Meade LX200 and SBIG ST7.

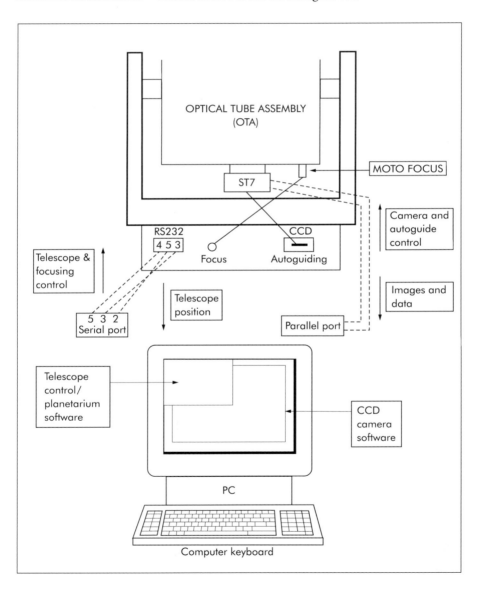

CCD cameras need to transmit a lot of data reliably and generally use the parallel printer port of the PC. The length of the cable is more critical here because the inter-core capacitance of the multi-cable and external electrical interference can degrade the image data. Starlight Express recommend a maximum cable length of 15 m for their cameras. I have used an SBIG ST7 on a 17 metre cable without problems and I know other ST7 users have used 25 or 30 metre lengths. But don't rely on it working. SBIG recommend the use of Beldan 8138 cables for extending the cable supplied with their ST7.

Because modern PCs running Windows 95 are proficient at multi-tasking, the remote observer can easily have both the planetarium/scope control software and the CCD imaging software running simultaneously on the same machine. The only disadvantage with this method is that many CCD camera packages don't let you re-size (shrink) the image window, so you may end up clicking between planetarium and imaging windows screens or having to put up with looking at a fraction of the CCD window if you have the planetarium window enabled.

This is no more than a minor annoyance, though. It is more problematic if one of your programs runs under DOS and one under Windows: it can sometimes prove impossible to operate both at the same time and a second PC may be required.

My own choice for controlling a Meade 12" LX200 and SBIG ST7 is to run the ST7 Windows (CCDOPSW) software alongside The Sky Level IV. This has worked well for me and many others. *Software Bisque*, who supply *The Sky*, also have available an extension to the software known as *T-Point Telescope Modelling* which enhances telescope pointing control. *The Sky* also offers the *CCDSoft* extension which enables CCD image-processing to be carried out alongside the standard planetarium package, instead of using two software packages.

Although users can experience considerable frustration with PC software from time to time, perseverance ultimately pays dividends and at least you can sort it out in the warm – you will be glad you made the effort when you can sit indoors on a freezing Winter's night and just reel in the images! This leap in technology is the major reason for the proliferation of amateur supernova and asteroid discoveries in recent years.

Advanced amateur astronomers now have the additional option of using professional astronomers' software because of the *Linux* operating system. Linux is a version of Unix, a powerful operating system used on mainframes and workstations, which can be used on a PC. In practice you can install Linux and use it in preference to, for example, *MS-DOS* or *MS Windows 95*. Once Linux is installed, powerful professional image-processing routines can be used. For more information, check out the Linux home page at http://www.linux.org/

Chapter 10

Video Astronomy

So far I have been talking about relatively long exposures with CCD cameras. In these applications, CCDs are being used at exposure times far beyond the 25 frames/second or 30 frames/second typically used for video camcorder applications. Even for our planetary CCD images we are usually considering exposures from 0.1 to 2 seconds, with an image scale of around 0.2″/pixel for 30–40 cm apertures. Nevertheless, amateur astronomers have successfully employed CCD cameras working at video frame rates to record the Moon and planets and even some of the brighter stars.

As far back as 1985 I employed an experimental industrial CCD camera to video-tape the lunar surface in "real time". The basic principles I described in earlier chapters of this book still apply, so if you want to record information at the resolution limit of your telescope you must ensure that at least two pixels span that smallest angle. This implies long focal lengths which in turn implies dim images at the CCD plane; and a dim image and short video-rate exposure do not go together well!

In practice, though, good results can be obtained if some compromises are made. CCD video-imaging of the lunar surface from half through to full Moon, or of Venus and Mars, is relatively easy with even modest reflectors. Video-taping Jupiter is a bit more borderline, but a scale of 0.5 arc-seconds per pixel will enable the best details visible on the majority of nights to be secured. However, Saturn really is a bit too faint for the highest resolution results to be achieved. An arc-second per pixel will produce an acceptable image of

Saturn with modern black-and-white security cameras, but this will not capture the finest detail visible.

Despite these limitations, CCD video work does have a number of advantages:

- The enormous number of images saved maximises the chances of capturing a few good frames when the seeing was momentarily good.
- The very short exposure time freezes shimmering detail for high-resolution images on the Moon and brighter planets.
- The video monitor enables rapid centring of objects in the field rather than having to "move and expose".
- Real-time events, such as lunar occultations of the brightest stars and planets, can be filmed; grazing occultations can be especially spectacular when a bright star winks on and off behind the lunar mountains.
- Total solar eclipses can be filmed and brought to life, capturing all the highlights almost forgotten by the visual observer at the time; the sound track of the exclamations enhances the memory!
- Coupling a CCD video camera to an image intensifier enables bright meteors and faint occultations (including asteroids occulting stars) to be recorded.

A standard domestic camcorder can be used to good effect with an amateur telescope, although some sort of mounting bracket is required (Figures 10.1 and 10.2, *opposite*).

Unlike an astronomical CCD camera, a camcorder has a lens permanently attached to the camera, to focus light onto the CCD chip. To enable your camcorder to image objects seen through the telescope you need to have an eyepiece in the drawtube, with the camcorder replacing the eye. In my own set-up, the final lens of the camcorder sits a centimetre or so away from the eyepiece lens. In the default wide-angle setting, the camcorder will automatically try to focus on the eyepiece body. However, by switching to maximum telephoto zoom, the view that the visual observer would see through the eyepiece will appear in the viewfinder.

As the images being viewed are often in the main black, and are subject to seeing effects, it is a good idea to focus the camcorder manually to prevent the autofocus continually hunting for the best focus position in the blackness. (There is an alternative to a domestic camcorder, a black-and-white CCD security camera,

Figure 10.1. The camcorder mounting on Gerald North's 46 cm Newtonian.

Figure 10.2. The simple camcorder mounting on the author's 49 cm Newtonian.

and this will often feature a detachable lens and a more sensitive, 0.1 lux, CCD chip for use in low lighting conditions. Such cameras are lightweight, inexpensive and easily attached to a telescope.)

Planetary observers will wish to search through all the best video sequences and freeze-frame the moments when the best seeing occurs. In addition, a permanent record of these best moments will invariably be required. Hard copies of CCD images are less desirable than digital images these days, when virtually every amateur owns a PC and can accept an image in GIF, TIF, JPEG, BMP or FITS standard. However, domestic camcorders and security cameras do not come fitted with PC interfaces.

Fortunately, inexpensive "frame-grabber" cards for PCs are widely available and, with these, video stills can be captured and saved to the PC's hard disk. Hard copies can then be printed (as above) via laserjet, inkjet or other printers. A video printer is an expensive option (but one which works with the Starlight Xpress frame store variant cameras too).

However, frame-grabbing or video freeze-framing tends to produce quite noisy results. An alternative, but more time-consuming method, recommended by some lunar and planetary observers, is to photograph the best moments on the video monitor with an exposure of a quarter of a second or so, in play mode, to avoid freeze-frame noise problems.

At lower image-scales, my own experiments in video astronomy have taught me that, in practice, through a 49 cm Newtonian, stars down to about mag. 6 or 7 can be imaged with camcorders and security cameras respectively. In fact, a camcorder on full zoom, without a telescope, can easily record 1st magnitude stars being occulted by the Moon.

Image Intensifiers

But what if we wish to image objects as faint as we can see with the dark adapted eye, in real time? This is where the image intensifier comes in. I have found that there is some confusion between CCDs and image intensifiers. The difference is this: a CCD chip collects photons, as electrons, in each pixel, for the duration of the exposure, which may be a tenth of a second or an hour. The accumulated charge held in each pixel is down-loaded at the end of each exposure and subsequently displayed on a monitor. An image intensifier, on the other hand, does not accumulate photons at all.

As each photon arrives it is converted into an electron, amplified as much as 100,000 times and immediately hurled onto a phosphor screen (which glows). The amplification factor sounds impressive, but the limiting magnitude is less so and is largely governed by the persistence of the phosphor screen (the length of time it takes an image to fade). In practice, and with a second-generation (micro-channel plate) image intensifier plugged into the telescope drawtube, the observer will see stars a magnitude or two fainter than might be seen through the eyepiece. For example, stars of mag. 17 might be glimpsed with a 40 cm reflector.

In practice, the few observers who do image-intensifier work use film holders that press photographic film onto the phosphor plate for a 1:1 "contact print". Unfortunately, due to the low resolution, electronic noise and short effective exposure times of these systems, the resulting image is not of the same quality as an equivalent CCD image.

Expensive commercial image intensifier systems are available in which a CCD camera is interfaced to the image intensifier via a fibre-optic bundle, but these are rarely available second-hand (which is how most amateurs acquire their image intensifiers).

The most effective applications of amateur image intensifiers are where real-time views, similar to those seen by the dark-adapted eye, are required, i.e. for faint occultation work and recording meteor showers. Andrew Elliott has been a pioneer of this type of work in the UK and has shown many dramatic videos at British Astronomical Association meetings (Figures 10.3 and 10.4, *overleaf*).

For video work it is essential to interface the video camera to the phosphor screen as efficiently as possible. Each point on a phosphor screen throws photons off in every direction; therefore, a lens placed as close as possible to the screen will capture the most photons.

A wide-angle camera lens and image intensifier, efficiently coupled to a video camera, can produce fascinating films of meteor showers, remarkably similar to the view seen by the naked-eye observer. Image intensifiers are available from suppliers who stock second-hand military equipment.

Amateur astronomers should be careful *not* to purchase "infra-red night vision" viewers which are now

Figure 10.3. The meteor intensifier system of Andrew Elliott: intensifier and lens on the left; video camera on the right.

Figure 10.4. Andrew Elliott's real-time telescope intensifier system (for occultation work): CCD video camera on the left; intensifier in the centre; electric focuser and telescope on the right.

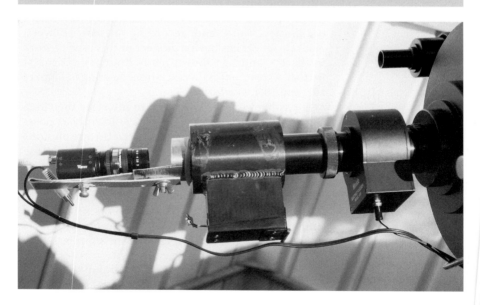

becomingly widely available. These viewers are very sensitive to infra-red light but they rely on the outdoor scene being illuminated by an infra-red LED mounted on the camera; they are nowhere near as sensitive as an image intensifier and are relatively insensitive to starlight.

Chapter 11

Observatories

An observatory may, to the beginner, seem like a secondary issue when contemplating the purchase of a modest telescope. Admittedly, a small telescope such as a 75 mm refractor or 150 mm reflector can be hauled out of the house whenever the clouds part, but lugging anything bigger outside can be a major effort.

It used to be thought that mounting a telescope on the roof of a building would be disastrous because the heat rising from the building would ruin the seeing. This has not prevented the leading planetary observers Miyazaki and Parker both having 40 cm f/6 Newtonians mounted on their homes!

The ease of use of a telescope/observatory plays a major psychological role in the observer's enthusiasm for astronomy and this factor mustn't be underestimated. From most parts of North America and Europe a clear night means a chilly night, except in midsummer. In midwinter, a clear night means a bone-chilling experience! In the UK, clear nights are a rarity and nights of exceptional clarity or steady seeing are extremely unusual, occurring maybe once every couple of months. All these factors and the natural aversion of human beings to the dark, damp and cold can seriously kill the enthusiasm of the most dedicated observer. Thus, a user-friendly telescope and observatory can be the vital factor that coaxes the amateur outside. A successful observatory design can certainly help create a successful observer.

The Simplest Solution

The simplest engineering solution open to the amateur is not to build an observatory at all. This means that the telescope has to be set up each time an observing session is envisaged. For smaller telescopes this is okay, but larger instruments pose more of a problem. In the US, city dwellers routinely lug substantial astronomical equipment in the backs of their cars and drive to dark sites as far afield as 100 miles from their homes! Equipment suppliers such as JMI can even sell you "wheeley bars" and other accessories to make telescope transportation easier.

In the UK, transporting telescopes to dark sites is a relatively uncommon pursuit. The UK has a high population density in the South, nor does the South have any truly high-altitude mountains. Really dark skies are therefore only attainable if you go to the most northerly English counties, Devon and Cornwall, Scotland, or parts of Wales. And the UK's extremely changeable weather means that guaranteed clear skies are a rarity – a succession of clear nights is usually the result of an unmoving high-pressure region settling over the country, itself a recipe for hazy skies.

Crystal-clear UK skies generally follow behind a fast-moving cold front travelling from North to South through the country, but the air following behind the front is invariably unstable, littered with shower clouds and often unpredictably brief. Transporting astronomical equipment any distance in the UK can be a thankless task.

Some leading UK observers manage to survive without an observatory. The renowned planetary observer Richard McKim uses a long-focus 30 cm f/7 Newtonian on a Dobsonian-type mounting. His telescope separates into alt–azimuth bearing and tube. The bearing sits on his patio and the tube lives in the garden shed. When good observing seems likely, Richard drags the tube out of the shed and positions it on the mounting, just outside his back door. Being primarily a planetary observer, he can then stroll in and out of his house waiting for moments of good seeing without running the risk of freezing to death.

Richard finds that he can memorise sufficient detail at the eyepiece to often make a sketch pad superfluous out-of-doors. This is a very simple, stress-free and enjoyable way of observing, but it does require a keen eye and experience at sketching fine details. Here is an example of a highly successful "observatory-free" system.

The successful British asteroid and supernova discoverer Stephen Laurie also observes without an observatory. He simply sets up his 25 cm Meade LX200 on the patio outside, on its tripod, and operates the telescope from indoors via the RS-232 data link and CCD camera parallel link.

Observatory-free systems like these can work well, especially if the instrument can easily be prepared for observing and is only a short distance from its shelter and from the house. Perhaps the most extreme (and successful) observatory-less observer is George Alcock. He has discovered both a comet (Iras–Araki–Alcock 1983d) and a Nova (Herculis 1991) using binoculars from inside his house and through double-glazed windows!

Simple Observatories

Many observers, especially those with large or sophisticated telescopes, prefer a permanent housing for their instrument but may not wish to go to the ultimate ideal of making a traditional-style dome.

Large and heavy instruments are often too cumbersome to be set up with ease and instruments which require polar aligning are best set up on a permanent pier. Moreover, a weather-proof environment outdoors is an ideal place in which to store auxiliary equipment such as computers and CCD cameras and in which to modify the telescope if adjustments are necessary.

Although domes are the traditional observatory buildings, they do have a number of disadvantages:

- They are difficult and expensive to build or purchase.
- They prevent the observer from being "under the stars" and seeing bright meteors etc. or cloud rolling in.
- Long exposures or slewing from object to object necessitate moving the dome slit with the telescope.
- Domes take a long time to cool down after a hot day, causing atmospheric turbulence.

The main *advantage* of a dome is the protection from the wind – especially bitterly cold, energy-sapping wind!

For my own three observatories I decided to forego the wind protection of a dome and settle for one of the simplest types of observatory: the run-off shed. A run-off shed is simply a shed on wheels that rolls over the telescope to protect it when not in use. Some run-off sheds are constructed from two halves which run off in opposite directions, with the telescope sitting near the join in the sheds. In this design, typified by the run-off shed used by Patrick Moore for his 12.5-inch Newtonian, the two shed sections are easier to move than one, single, heavy shed. The disadvantage of the two-ended run-off shed is the join; the shed should be water-proof, and guaranteeing a water-tight joint can be very difficult.

In the single-ended shed, which I prefer, water-proofing is less of a problem. But the single-ended shed needs to be rigidly designed as the missing side can cause considerable flexure of the shed when it is being rolled back. A shed that is too massive can be very tiring to roll back, especially when the observer is tired and cold in the early hours.

The removable side of the shed can be designed in a number of ways. My first run-off shed, for my 36 cm Cassegrain–Newtonian (Figure 11.1), featured two "half-sides" with handles. These were totally removed when the telescope was in operation and were located with pins and hinged fasteners on the shed side. After some fifteen years of this design I tired of the effort of removing these heavy sides and they were replaced with a double-hinged door which was far more successful.

Figure 11.1.
Run-off shed for the author's 36 cm Cass–Newtonian.

My first wheel-rail design was naïve beyond belief: simple nylon wheels on the base of the shed ran (largely by luck) on L-section angle-iron which stretched between wooden posts hammered into the grass lawn. After several disastrous nights rescuing this shed, late in 1980, I screwed the L-section angle-iron rails firmly to wooden timbers along their entire length. Finally I switched to U-section rails. The U-section channel is generally reliable, although if a stone gets stuck between the wheel and the U-section it can cause the shed to jam. Brushing the U-section rails of this shed was a regular chore! The problems I experienced with my first run-off shed led to a complete rethink for my second shed (for a 0.49 m f/4.5 Newtonian), as shown in Figure 11.2.

The most reliable wheel-rail solution is really quite obvious; use a deep V-groove pulley wheel and an inverted T-section rail. Such rails are easily obtainable in long lengths from scrap metal yards and are easily screwed down to timber supports. A V-groove pulley wheel restricts the shed's sideways motion and the shed can never jump off the rail. Pulley wheel units can easily be obtained from most hardware stores. An additional advantage of this design is that leaves and stones cannot cause any problems. The only aspect of this solution which does need special care is ensuring

Figure 11.2.
Run-off shed for the author's 49 cm Newtonian.

that the inter-rail distance is precisely right and precisely the same along the length of the rails. This can
easily be achieved by delaying the final "screwing
down" of the rails until the completed shed is running
freely.

All my run-off sheds are prevented from rolling
about on their rails by turnbuckles anchored to the
inside of the timber rail supports and attached to
hooks on the inside of the shed base. A total of four
turnbuckles are employed, two on each rail. Dust and
dirt entering the shed (between rail and shed wall base)
are minimised by a plastic skirt all around the base of
the shed.

For my third run-off shed (Figure 11.3) I decided to
construct an observatory which was as user-friendly as
possible, quick to set into operation and hassle-free.
The shed was designed around a 30 cm Meade LX200,
equatorially mounted on a concrete plinth and at the
optimum height for easily observing objects at virtually
any altitude. With my other two run-off sheds, the
sheer weight of the shed and the telescopes had proved
a major deterrent to going out and observing, especially on nights of indifferent weather.

Figure 11.3.
Run-off shed for 30 cm
LX200.

Unless you're very fit it is all too easy, in the cold and damp, to strain a muscle while moving heavy telescopes around the sky and moving heavy sheds along their rails. Large Newtonians often necessitate "step-ladder" observing positions with very little observer comfort. This is an aspect of observing that the beginner often overlooks when choosing a telescope or observatory.

Finally, I have experienced *many* astrophotography/ CCD imaging sessions which have seemed to be nothing less than a battle to the death against defective equipment, the dark, damp and cold, and the rapidly changing British weather. There are a near-infinite number of conspiring factors which can prevent you from having a good observing session; building a friendly and ergonomic observatory can definitely lessen these bad experiences.

If It *Can* Go Wrong

To digress slightly and list just a few of the disasters that have befallen me may be instructive; I have suffered:

- Damp conditions affecting the mains power supply, ultimately tripping the house mains off.
- Defective CCD cameras, computers, intermittent electrical connectors and telescope drives.
- My eyelash freezing to the eyepiece.
- Misted or frozen-up CCD faceplate windows.
- Dead illuminated-reticle batteries.
- Neighbours' bonfire smoke drifting across the field of view and obscuring Comet Hale–Bopp on one of the clearest of nights.
- Dewed-up main mirror, secondary mirror or guidescope lens.
- Seriously gouging my head on the 0.49 m mounting while searching for an electrical lead on the ground. This observing session was terminated owing to the volume of blood spraying out of the wound!
- Falling off an observing step-ladder onto a prickly hedge and thence off the hedge into a muddy field.
- Dropping a Nagler eyepiece onto a concrete floor.
- A toggle switch on my framestore CCD camera that snapped off during an observing session.

- Scratched film due to "out of doors" film-loading in a dirty environment.
- Finding that the optics had shifted during an exposure, so the photograph was hopelessly blurred.
- The run-off shed getting stuck in ice which had formed on the rails.
- Specks of dust on the CCD faceplate ruining planetary images.
- Walking into washing-lines and tree branches.
- Slipping on ice in the darkness.
- Finding that the eyepiece position was impossibly inconvenient to make the observation.
- Totally failing to find/recognise the field.
- Guiding meticulously for 30 minutes on a variable-star field which I had been told was at *minus* one degree in Declination and it was actually at *plus* one degree.
- Finding that the most recently discovered comets are *always* in the parts of the sky obstructed by trees and neighbours' houses.
- Mistaking the position of a comet or the time of an event by stupidly confusing the date of an early morning event or mistaking UT for BST, or vice-versa, in the Spring and Autumn.
- Time after time, being clouded out only seconds before starting an exposure after half an hour or an hour of setting the equipment up.

This list of tragedies that have befallen me is far from exhaustive, but it does at least afford a little amusement (and dare I say instruction) for others, and it reinforces the need for a well-thought-out observatory, such as that for my third main instrument.

Although amateur astronomy is a fairly safe hobby (with few fatalities!), sensible electrical precautions should be taken when working outside in the dark and damp. The easiest step to take is to buy an *earth-leakage circuit breaker* (ELCB) from which the observatory can be powered. These inexpensive devices detect current flowing to earth (e.g. via a human being!) and cut the power within milliseconds of a potentially hazardous situation developing. It is also sensible practice to check that mains cables used outdoors do not become frayed and are not left in positions where they frequently become damp.

The advent of commercial computer-controlled fast-slewing telescopes also opens up the highly desirable option of remote observing, in which the observer sits indoors in a warm room and moves the telescope around from the computer keyboard while watching the images arrive via the CCD camera. My third run-off shed was designed with remote operation in mind; cables to control the LX200 (see Figure 11.4) and auto-guiding SBIG ST7 CCD camera were laid under the garden lawn for a distance of 15 metres. This was the maximum distance that I wanted to run the CCD camera cables for fear of degrading the image quality. As I have already mentioned, a remote controlled auto-slewing telescope which is going to be used for super-nova patrolling or monitoring numerous objects each night needs to be in the open air – or the dome needs to be computer-controlled too!

A run-off shed is only one solution to this problem; the most popular alternative, without resorting to a dome, is the run-off roof observatory. As its name suggests, the run-off roof observatory is a building which

Figure 11.4.
Underground cable laying for LX200 remote control.

allows the roof to slide back to allow the telescope access to the night sky. This design (Figure 11.5) has the advantage over the run-off roof shed in that a moving roof is much lighter than a moving shed. The height of the stationary walls is a major design consideration with the run-off roof system. High walls afford some protection for the observer from biting cold winds, but they may well prevent the telescope from reaching low altitude objects. Compact, fork-mounted telescopes, such as Schmidt–Cassegrains, are well suited to this design as the telescope altitude above ground varies little with the orientation of the telescope.

Shed-like observatories are undoubtedly versatile and relatively easy to construct; however, many observers prefer the civilised ambience, protection from winds and freedom from dew that only a dome can bring. So let us now have a look at the structure that everyone always associates with an astronomer, the observatory dome.

Figure 11.5. Mark Armstrong's run-off roof observatory.

Observatory Domes

My first experience with an observatory dome was as a twelve year old. My physics teacher was interested in astronomy and, luckily for me, had the keys to an observatory dome mounted on top of a building known as the Athenaeum, in Bury St Edmunds. The building housed a splendid, brass-tubed, Victorian 10-cm refractor, and every few months I was able to use the telescope alongside friends from the school astronomy club (which I had formed). Although I dislike domes from the viewpoint of observing efficiency (not seeing the clouds coming, losing orientation with the sky etc.), I have to admit that being inside a small observatory dome makes me feel like I have been transported back to a more civilised age when the planetary observer was king and brass-tubed refractors were a status symbol. Domes have a certain distinction that other observing structures simply do not have.

Having said this, they can also be intimidating buildings, especially if the dome is hard to move and the slit is narrow, necessitating frequent movements to allow the observed object to stay in view. I well remember a night spent observing the first quarter Moon with the Greenwich 28-inch refractor in the mid 1980s. It was certainly an experience, but the effort involved in rotating the dome was considerable.

A good book to consult is Patrick Moore's *Small Astronomical Observatories*, published in Springer's *Practical Astronomy* series. Twenty-four observatories are described in detail, together with detailed observing plans; inspiration for the budding observatory builder!

Observatory domes can be purchased as a complete unit, but good ones are prohibitively expensive. One of the world's leading suppliers of quality domes is Ash Manufacturing Co. of Plainfield, Illinois, USA. As with most things in this world, if you want quality, you have to pay for it. Ash domes are made from galvanised steel and are in use at many University observatories throughout the UK and USA.

UK manufacturers of domes have come and gone over the years. Most offered a light weight fibreglass construction with a sliding "up and over" shutter. These can become very hot and stuffy in the summer, whereas metallic domes tend to reflect the sunlight more efficiently.

Some consideration of the survivability of a dome in gale-force winds is essential. Many domes simply sit on their observatory walls by virtue of their weight, and if the dome is of fibreglass or stretched fabric construction, strong winds can cause big problems. The most off-putting aspect of the dome from the constructors viewpoint is the construction of a hemi-spherical surface and the circular running track on which it rotates. The dome slit is also an intimidating aspect of the design. However, many amateur astronomers with basic wood-working skills or just plain ingenuity have solved these problems.

When you look at the domed amateur observatories around, there appear to be five distinct categories of dome:

- The converted grain silo top dome.
- The metal skeleton, covered with aluminium sheeting.
- The wooden skeleton, covered with aluminium sheeting.
- The all wood (often hexagonal/octagonal) dome.
- The fibreglass dome.

The converted grain silo top is very popular among amateurs who have a very large instrument to house. Denis Buczynski houses his 53 cm Newtonian in such a dome (Figure 11.6). However, most silo tops are at least 5 metres in diameter, a bit excessive for a 20 cm Schmidt–Cassegrain!

The metal or wooden skeleton approach is, perhaps, the best prospect for the novice wood or metal-worker. In both designs it is important to get the dome base ring, on which the dome rotates, as circular as possible to avoid rotation problems in the final stages. Professional machine-shops or carpenters can easily produce an accurate base ring if the amateur feels this is beyond his abilities. Once a base ring has been established, the remaining metal or wood skeleton is prepared by producing circular hoops on which the skin of the observatory and the

Figure 11.6. Denis Buczynski's converted silo-top dome housing a 53 cm Newtonian.

shutter are supported. A jig-saw and plywood are employed in the wooden variant; metal hoops, bent to size, will do nicely for the metal skeleton. Once the skeleton has been constructed, thin sheet aluminium can be screwed or riveted to the skeletal structure (Figure 11.7).

Fibreglass (glass fibre) domes are a popular alternative, especially among those workers who have used fibreglass before. Fibreglass construction technique involves first preparing a mould onto which glass fibre matting is positioned. This is then impregnated with resin and allowed to set. In practice it is impractical for most people to produce a mould for the whole dome, so a mould for a dome segment is produced and the segments are then joined together to form the dome.

Whichever style of domed observatory you choose, a decision needs to be made about the type of dome-slit which is incorporated.

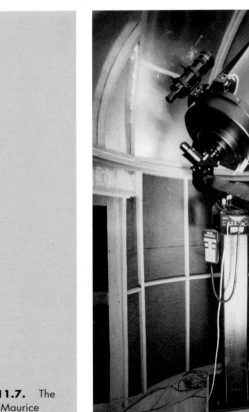

Figure 11.7. The inside of Maurice Gavin's wood and aluminium dome.

Many observers, in retrospect, wish they had made their dome-slit wider when they built their observatory. A wide slit provides more time for the telescope to track an object, allows the inside of the dome to cool more rapidly and enables more of the sky to be seen by the observer. The main disadvantage of a wider slit is that it weakens the overall structure of the dome, but this can be compensated for by ensuring that the dome skeleton is rigid. It is essential that the shutter enables the zenith to be easily accessed, so the gap in the dome must extend well beyond the dome top. Objects near the zenith are at their very best and there are few things more frustrating than not being able to observe an object because the dome roof is in the way! Similarly, the telescope should be mounted in the dome such that it can reach objects close to the horizon, which means that the base of the dome-slit should not be higher than the telescope's declination axis centre-point.

I don't intend going into observatory design in any greater detail, as there is already a book in this series devoted to the subject. Suffice to say, every amateur will have a unique set of skills and materials to hand and, with a little patience and ingenuity, can easily construct a practical observatory to his taste (see Figure 11.8).

Figure 11.8.
Cdr Henry Hatfield's wooden beehive observatory.

One final word of warning. Some amateur astronomers have decided to move house and inadvertently sold the observatory building and contents to the next purchaser! Any buildings in a garden, and the contents of those buildings, can be lost if you don't specify that they are *not* included in the sale of your house!

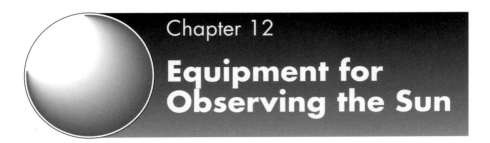

Chapter 12

Equipment for Observing the Sun

Astronomy is at least a relatively safe hobby. However, there is opportunity for the observer to seriously injure himself. I have already mentioned the precautions required for using electricity in damp conditions, and being aware of the dangers that the unseen tree-branch can pose. However, if there is one field of observation where caution is always necessary, it is in the field of solar observation. Looking at the Sun through any telescope, with filters or not, is a highly dangerous pursuit unless you know *precisely* what you are doing.

Just because the Sun may look dim enough to observe through a filter does *not* mean that it is. Infra-red radiation is invisible to the eye, and you can easily be lulled into a false sense of security because an image of the Sun doesn't *look* dazzling.

Many amateurs, including experienced observers, have looked through solar filters at a pleasant solar image, unaware of the large amounts of infra-red or ultra-violet energy still reaching their eyes. The result, in some cases, has been PERMANENT BLINDNESS.

Amazingly, manufacturers *still* advertise and sell solar filters that are unsafe. Indeed, the vast majority of solar filters sold in the high street stores with cheap small refractors are *unsafe*. Remember, infra-red radiation (and damaging ultra-violet) light is invisible, and A DIM IMAGE IS NOT NECESSARILY A SAFE IMAGE.

Eyepiece Projection

One of the safest ways to observe the Sun in white light is to project the image onto a card using eyepiece projection (Figure 12.1). When using a refractor, a large cardboard disk slipped over the telescope tube can be employed to create a shadow, thus enabling the projected image to be viewed in broad daylight. A sketch of the projected image can then be made, or the image on the card can even be photographed.

Note that some manufacturers (for example, Meade, in the manual for the ETX) strongly advise *against* using this method, because it can damage the telescope. If the solar image drifts off the centre aluminised spot on the corrector plate of a Schmidt–Maksutov it can fall on the upper part of the tube. If this happens, it will heat up the top of the telescope tube and can even burn it, or in extreme cases damage the corrector plate itself.

Direct Solar Observation

A small percentage of solar observers – the most experienced – observe or even photograph the Sun directly. I would certainly not recommend this to the beginner or anyone with a tendency towards impatience. Advanced

Figure 12.1.
Projecting the solar image.

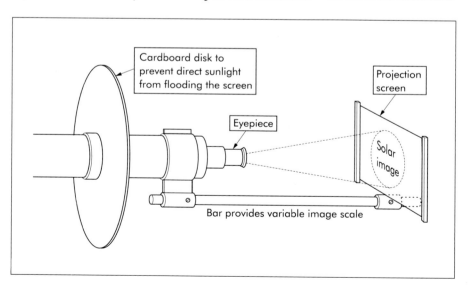

Cardboard disk to prevent direct sunlight from flooding the screen

Projection screen

Eyepiece

Solar image

Bar provides variable image scale

solar-observing requires an understanding of the problems involved, and meticulous and painstaking preparation and planning. Otherwise a disaster may result, and our eyes are the most precious possessions we astronomers have.

For those observers who become bitten by the desire to obtain stunning solar photographs, the next section is for you, but I accept no responsibility for anyone who goes down this path without taking the necessary precautions.

Observing in White Light

Solar Filters: Inconel

The simplest form of advanced solar work involves direct observing or photography of the Sun through the telescope. Don't skimp, buy one of the best solar filters. *Thousand Oaks Optical* of Thousand Oaks, California are generally regarded as the best solar filter manufacturers. Their solar filters are made from polished glass coated with stainless steel and nickel–chromium alloy (inconel) which allows a tiny fraction of sunlight to pass through the filter, reflecting the rest safely away from the telescope.

In the world of filters you will hear the term *neutral density* or "ND" used. This is usually followed by a number. Neutral density simply means that the filter is neutral in terms of colour bias, and attenuates visible light of all colours by the same amount. Thus the Sun will appear white or slightly yellowish when viewed through the filter. The number indicates the power of 10 to which the filter attenuates, i.e. a filter with a ND of 4 attenuates by 10^4 or 10,000 times. Thousand Oaks' visual filters (types 1 and 2) attenuate by 100,000 fold and their photographic filters by 10,000 fold (to ensure short exposure times with a bright image).

Solar Filters: Mylar

You will see advertisements for metallised mylar film, which some observers use as solar filters. Although this type of filter has been used successfully by many amateurs, I strongly discourage its use unless the observer has scientific proof of its transmission characteristics. Mass-produced mylar film can also suffer from surface imperfections in the form of pinholes and the surface can easily be damaged, especially if the mylar is only metallised on one side.

Mylar also has the disadvantage that it polarises light passing through it, producing spurious "wings" on the solar image and making it unsuitable for photography. Visually, the Sun appears blue when seen through a mylar filter.

I personally recommend avoiding mylar film for solar observing: you get what you pay for and your eyesight is not worth a philosophy of false economy!

Seeing

As with planetary photography, atmospheric seeing is crucial to high-resolution solar work. Unfortunately, when the Sun is above the horizon the Earth's atmosphere is very turbulent, far more so than in the middle of the night. Thus it is essential to keep exposure times to a minimum to "freeze" moments of good seeing. Fortunately, the amount of sunlight available when using a ND4 filter is more than enough to guarantee very short exposure times, typically $1/100^{th}$ to $1/1,000^{th}$ of a second. Again, sensible precautions and a high level of common sense are essential when photographing the Sun. Poor-quality filters can easily send enough infra-red or ultra-violet radiation through the viewfinder of an SLR camera to damage the eye.

Small-aperture refractors are the favourite instruments for solar observers, whether they are using the projection method, or visual/photographic observing methods. Because of poor daytime seeing conditions, large apertures are rarely necessary. A 60 mm refractor can often resolve all the detail that is visible. Indeed, apertures greater than about 12 centimetres are undesirable, owing to the amount of heat being dissipated within the telescope tube. In addition, the closed tube of a refractor will generally yield better daytime resolution than a reflector and scattered daylight will also be less of a problem.

Hydrogen-Alpha Equipment

If you really get addicted to solar observing you may wish to view or photograph the Sun's Hydrogen-alpha (H-α) activity and look at the surface features and spectacular prominences (see Eric Strach's Figures 12.2 and 12.3).

As all amateur astronomers will know, during a total solar eclipse the Moon's disk covers the Sun and the solar corona and prominences are revealed. The

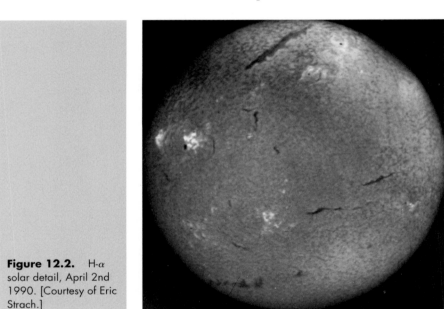

Figure 12.2. H-α solar detail, April 2nd 1990. [Courtesy of Eric Strach.]

Figure 12.3. Loop prominence in H-α, August 17th 1989 [Courtesy of Eric Strach.]

corona is very faint and, in practice, unobservable by simple optical methods, except during totality. However, modern technology enables us to use filters to view the solar prominences without having to wait for the next eclipse. Subtle H-α details on the surface can also be viewed; these details are different from the sunspots and faculae that are observable in white light.

The key to observing these solar phenomena is the ability to fabricate extremely narrow-band filters centred on the wavelength of Hydrogen-alpha, i.e. 656.3 nm, at the far-red end of the visual spectrum (with the exception of the spectrohelioscope).

The technology required to construct affordable H-α filters (and the rarer Calcium line filters) simply did not exist before the 1970s and the pioneering work by the Daystar corporation. Solar prominences and associated phenomena emit light at the H-α wavelength. Unfortunately, the solar disk is considerably brighter than these phenomena in white light and so is the Earth's atmosphere which, as we all know, appears extremely bright in the direction of the Sun (although a tolerably pleasant blue only tens of degrees away). A narrow-band H-α filter rejects the vast majority of the solar radiation and lets only wavelengths around the H-α band through. Thus, the prominences and subtle surface details are no longer swamped by the rest of the spectrum and the bright sky. Being in the far-red end of the visual spectrum, CCD cameras/security video cameras are well suited to imaging H-α phenomena.

H-α filters should be used in conjunction with Energy Rejection Filters, or ERFs. An ERF filters out the harmful infra-red and ultra-violet radiation before it gets to the H-α filter. Although the H-α filter can filter them out, it is much more sensible to use an ERF in front of the telescope objective. At this point the radiation has not entered the telescope, is not focused and concentrated, and the ERF will protect the very expensive H-α filter from the concentrated radiation that it that would otherwise fall on it.

H-α filters range in price from hundreds to thousands of dollars, so prolonging their working life with a relatively inexpensive ERF is a wise precaution! Incidentally, H-α filters can only filter correctly if the beam of light is f/30 or slower: a 50 mm refractor of at least 1,500 mm focal length would be appropriate. Large apertures can always be stopped down to f/30. As you might expect, the wider the passband of the H-α filter, the cheaper the filter.

For many years the US astronomy supplier Lumicon has marketed its own, relatively cheap, H-α filter. This filter has a bandpass of 1.5 Å (Å = Ångstrom = a tenth of a nanometre) which is enough to show the Sun's prominences but little H-α detail on the disk. However, this filter has allowed many amateurs to obtain goodish views of the solar prominences. The filter can be tilted within its housing to fine-tune it on the H-α line. (Incidentally, solar activity follows an 11 year cycle; the last maximum was in 1990/91 and the next will be around 2001/02, so activity will be increasing nicely by the time this book gets into print in late 1998.)

The Daystar corporation, and its founder Del Woods, have always been a top supplier of high-quality sub-Ångstrom interference filters. With bandpasses of 0.5 Å, 0.6 Å, 0.7 Å and 0.8 Å, not only are the solar prominences visible, but so are fine surface details like solar flares and spicules. Early Daystar filters were primarily of the tilting type, i.e. fine-tuning by tilting the filter in the beam. Nowadays the majority are tuned by temperature regulation. In recent years, the eyepiece and apochromat company Tele Vue have teamed up with Daystar to produce telescopes for H-α observing.

Tele Vue's *Solaris* telescope comprises a 60 mm short-tubed refractor with an effective f/30 focal length (the objective lens is f/9, but a special corrector lens/Barlow boosts this to f/30). A 60 mm ERF is supplied as standard, along with a 0.7 Å Daystar filter. The result is an extremely compact, portable, but expensive, solar observatory. It is ideal for taking on eclipse trips to check out the prominences *before* totality. A kit for converting the 100 mm aperture f/5 Tele Vue Genesis into an f/30 H-α system is also available.

Tele Vue are not the only company offering solar kits for their refractors; Astrophysics sell kits for their apochromats too. On this point, because solar interference filters operate over such narrow bandwidths, the achromatic quality of the refractor used is irrelevant. Focusing light of a single wavelength is the aim, not across the visual spectrum.

Another piece of equipment that the die-hard solar observer may be interested in is the *Baader Planetarium H-α Prominence Viewer*. This device is modelled on the early coronograph/prominence scopes (Figure 12.4, *overleaf*) which were in the hands of a few amateurs in the 1970s, before the sub-Ångstrom H-α filters became affordable. They do use filters, but at 4 and 10 Å they are totally incapable of showing any H-α surface detail.

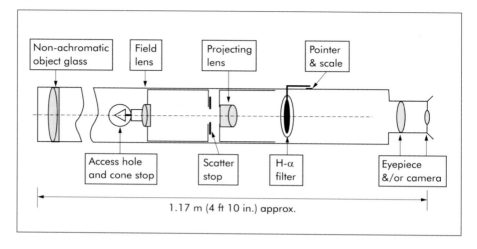

| Non-achromatic object glass | Field lens | Projecting lens | Pointer & scale |
| Access hole and cone stop | Scatter stop | H-α filter | Eyepiece &/or camera |

1.17 m (4 ft 10 in.) approx.

Figure 12.4.
Diagrammatic layout of a prominence telescope.

Indeed, with such a wide bandpass the solar disk would be intolerably bright in the telescope eyepiece were it not for the fact that a metal occulting disk, placed at the telescope's focal plane, is used to shield the solar disk from view. The Baader Planetarium viewer resembles a huge Barlow lens and contains the occulting disk and a series of lenses and diaphragms to project the focal plane image of the occulting disk and solar prominences to the eyepiece/camera with the minimum transmission of glare and stray light. The view through the instrument is remarkably similar to that seen at totality during a solar eclipse, albeit without the spectacular corona. The extra light transmission of the 4 and 10 Å filters is a big advantage to the prominence photographer who always wants a bright image for a shorter exposure time. Indeed, if prominence photography is your sole aim and you live in a region with less-than-perfect seeing (don't we all!), then a Baader Prominence viewer may be a better purchase than a sub-Ångstrom filter.

But this is not the only consideration. The Baader device is quite tricky to use and the solar disk is uncomfortably bright when it is not hidden behind the occulting disk. Of course, a polar-aligned telescope with a good equatorial drive is essential for keeping the image behind the disk. Also, the observer needs to move the image behind the disk when the observing session starts. Even advanced solar observers who know what they are doing feel uneasy about this part of the operation; at all times the safety warnings supplied with the Baader viewer must be observed. Fortunately, the price of this instrument and the fact that it is only purchased by solar observation experts means that serious eye injuries are unlikely, but – once again – you have been warned!

If you want an easy and comfortable-to-use H-α system, I recommend the Tele Vue Solaris in preference to the Baader Prominence viewer.

The cheapest Baader viewers are designed for use with Vixen and Celestron's standard refractor focal lengths (the occulting disk(s) need to be precisely machined to match the telescope's focal length and the diameter of the resulting solar image). Custom viewers for non-standard focal lengths can be made to order.

H-α equipment is expensive and for the specialist; however it is interesting to note that many people spend as much money travelling to see a total solar eclipse as they could spend on a premium H-α filter.

Viewing Eclipses

Those (like me) who have become almost addicted to eclipse-chasing may well want to consider equipment suitable for taking to an eclipse.

Most amateurs going to a total solar eclipse want to bring back a souvenir of the experience, usually in the form of a photograph during totality, or of the diamond ring. While this is a natural desire, once you have been on an eclipse trip you may question whether photography is worthwhile. The point here is that total solar eclipses are short events. The longest events possible are only seven minutes long and many are much shorter, typically 1–4 minutes. With such a short event it is essential to savour the visual experience; to end up, as some people do, spending most of totality fiddling about with camera equipment is a tragic waste of the time that you will spend in the Moon's shadow. There are so many things to savour: the Moon's shadow coming and going, the bright stars and planets appearing, the prominences on one edge disappearing and those on the opposite edge sliding into view. It is a sad fact that most people's solar eclipse photographs are far worse than they hoped, primarily because of shutter vibration. Eclipse photographers typically use long telephoto lenses on shaky tripods and exposures far shorter than they would normally use: the result is a blurred mess.

I can only offer my own strategy here, based on personal experience. I now take just a couple of frames during mid-totality; these are 1-or-2 second exposures on Fuji Velvia slide film (50ASA) with a 90 mm aperture f/11 Celestron spotting scope (see Figure 12.5). Mid eclipse is the quietest time during totality, with most of the prominences hidden and the observers mid-way between the frantic second and third contact times; so you are unlikely to miss anything dramatic by taking a few quick photos.

But I now video the eclipse as well, and the video is a hundred times more exciting than the photographs can ever be. Modern camcorders are extremely sensitive to light and extremely tolerant to changing light levels.

I would recommend purchasing a basic 5× teleconverter for your camcorder and carrying out trials on the full Moon in the months leading up to the eclipse. Using the 5× teleconverter screwed onto the camcorder lens, adjust the camera zoom and converter focusing barrel until the full Moon fills about a quarter to a third of the screen. The Sun/Moon will take 2 minutes to drift by their own diameters, so an equatorial drive is not essential (and is cumbersome to take abroad). This sort of magnification yields a superb result on my camcorder (Panasonic NV-S90B), with the

Figure 12.5. Total solar eclipse, November 3rd 1994: Celestron C90; 1000 mm f/11; 2 sec on Fuji Velvia. [M. Mobberley.]

corona not being overexposed (as can happen at lower powers) and the camcorder adapting perfectly to the lighting conditions (see Figure 12.6). Modern camcorders can also cope with the dazzling light 30 seconds before and after second and third contact, so you can whip the solar filters off the camcorder 30 seconds before totality starts and 30 seconds after it ends.

This is the one situation where I actually do tolerate mylar filters – to enable the positioning and focusing of camera and camcorder prior to totality. Even so, I always make sure the mylar is from a reputable dealer and minimise the time spent looking through the camera viewfinder. Mylar filters should be inspected for pinholes prior to use. If pinholes are present they should not be used.

Most people's memories of precisely what happened during a total solar eclipse are fairly confused, even 10 minutes after the event. A good video brings the experience back to life and also captures the gasps of awe from the observer and his or her friends. A camcorder can also be left pretty much to itself, apart from

Figure 12.6.
The author's eclipse camcorder and Celestron C90 mounting.

occasional adjustment as the Sun/Moon drifts and replacing the filter at the end. A film camera cannot be left to itself and I would strongly resist the urge to use a command back/motor power wind on a film camera. Telephoto lenses on tripods and long exposure times will suffer enough from camera shake without any help from a motor drive. This is why I now stick to 1-second exposures of the corona: most of the vibrations die down in the first quarter-second or so, and the Sun/Moon will only move by 15″ during a second, so slow film and a long exposure can pay off. If you really want sharp photos of the diamond ring/prominences, then you will need to increase the stability of your system or use faster film/shorter exposures to get faster than about 1/60th of a second.

My choice of a 1,000 mm system for eclipse photography is largely governed by my desire to purchase a 1,000 mm spotting scope which can be used for more than just eclipse photography; many will prefer 500 mm or 600 mm telephoto lenses so that several degrees of solar corona can be accommodated.

Despite my remarks about savouring the visual experience, a few eclipse-chasers make it their goal to take excellent eclipse photographs, especially those showing as much intricate coronal detail as possible. The main challenge here is to capture the full dynamic range of the event, and reproduce the view seen by the eye. This is a real problem as the solar corona is very bright near to the Sun's surface, but elusive and faint a degree or two away. To counteract this, a few amateurs have designed ingenious systems to try to reduce the light from the inner corona while preserving the light from the outer. The standard trick for achieving this is to suspend a circular disk between the lens and the film plane which cuts off the light near the Sun's surface by a factor of a hundred or so, while allowing all the light to pass to the outer edges of the film (Figure 12.7, *opposite*).

Because the disk is above the film plane, it appears out of focus and thus does not cause an abrupt change in brightness by its shadow. Nevertheless, the disk must be centred precisely above the image of the Sun on the film and so rigorous techniques have to be applied in building and using such equipment. Dr Francisco Diego of University College, London has refined this technique in recent years.

An alternative option is to take a variety of exposures of the eclipsed Sun, either with a film camera, or a

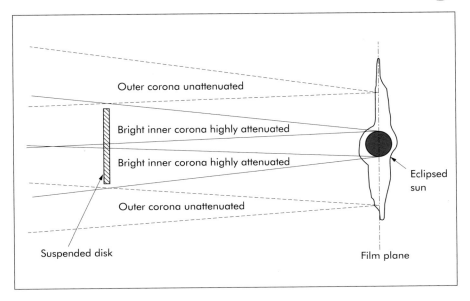

Outer corona unattenuated

Bright inner corona highly attenuated

Bright inner corona highly attenuated

Outer corona unattenuated

Suspended disk

Eclipsed sun

Film plane

Figure 12.7.
Attenuating the corona by suspending an afocal disk in the light path.

digital camera, and then combine the inner coronal (short) exposures with the outer coronal (long) exposures using a PC and software such as *Adobe Photoshop*.

However, these are techniques for the specialist; my own advice remains the same: video the eclipse or even take a few snapshots, but, above all, enjoy the visual spectacle!

Chapter 13

Star Atlases and Deep Sky Atlases

It might be thought, in this computerised multimedia age, that the traditional star atlas was a thing of the past. However, I am pleased to report that this is not the case. Despite the advances in PC monitor resolution and performance of recent years, everyone agrees there is still no substitute for the traditional paper book or star atlas.

There are now four Star Atlases that stand head and shoulders above the rest: *Norton's 2000 Star Atlas and Reference Handbook, Sky Atlas 2000, Uranometria 2000.0* and the *Millennium Star Atlas.* Norton's Star Atlas is published by Longman/John Wiley and the other three are published by Sky Publishing.

Norton's 2000 Star Atlas first appeared in 1910. Generations of astronomers have used it and it is the best-known star atlas in the world. The eighteenth edition, revised by Ian Ridpath in 1989, is still available and was the first edition to incorporate 2000 epoch co-ordinates. Norton's is more than just a star atlas. The reference section contains information on time, practical astronomy, the solar system, stars and galaxies, and numerous useful tables. It's a must for every astronomers' bookshelf. The star charts essentially show the sky to the limit of the dark-adapted eye from a really dark country site, i.e. magnitude 6.

Sky Atlas 2000 reaches to magnitude 8 and the chart scale is about 8 mm to one degree, roughly twice as generous as Norton's. This is a convenient atlas to take out of doors, especially as a laminated version is available. This is an excellent atlas for the naked-eye and small-binocular user. It is also relatively inexpensive and, like me, you will probably end up scribbling pencil marks all over it. Indeed, with every atlas I have, I started out wanting to preserve it in pristine condition but I have ended up with scrawls like the trail of a demented spider over all of them as new novae and comets were discovered. *Sky Atlas 2000* comes in a desk (black stars on a white background) and field (white stars on a black background) edition. The field edition is extremely useful under the stars.

Moving up to the two-volume set of *Uranometria 2000.0* (Figure 13.1), we are now drifting into the realm of reference star catalogues rather than star charts for outside use. This set covers both hemispheres to magnitude 9.5 and was absolutely invaluable to me when I was carrying out a nova patrol with 50 and 85 mm lenses. The magnitude limit of my 1-minute exposures was so similar to that of Uranometria that it made the checking of suspects a very easy business. A huge number of galaxies, nebulae and variable stars are

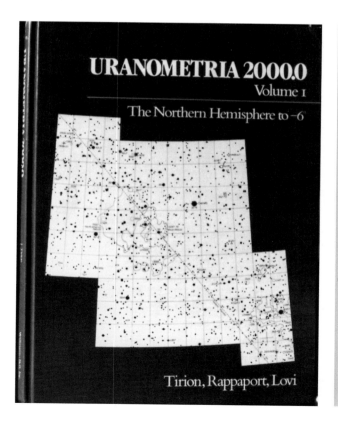

Figure 13.1.
Uranometria 2000.

marked, and photographer or binocular user, you will find the atlas indispensable.

The Millennium Star Atlas, published by Sky Publishing, is the most impressive star atlas available to the amateur. The three-volume set is massive and the $250 price (1998) reflects the investment. Each volume contains 516 star charts, 235×330 millimetres in size. The entire sky is covered down to magnitude 11, i.e. more than 1,058,000 stars are plotted. In addition, 8,000 galaxies, 900 nebulae, 860 star clusters and 9,000 variable stars are shown. The latest data from the Hipparchos satellite/Tycho catalogue is incorporated and a generous scale of 100″ per mm is adopted throughout. This, incidentally, corresponds to a focal length of 2060 mm/81 inches, so if your telescope is around this focal length you can lay your photographic negatives on the pages to check the field! No dedicated amateur should be without this superb atlas.

Star atlases are not the only paper atlases that the amateur will want to have. Despite the proliferation of multi-media astronomy CD-ROMs, a quality book of galaxy images still looks pretty impressive on the observer's bookshelf. A short list of the best galaxy and miscellaneous atlases (especially useful for the supernova patroller) appears below.

The Carnegie Atlas of Galaxies (two volumes) by Sandage & Bedke (1994). Carnegie Institute of Washington DC. 1,200 photographs of galaxies from Palomar, Las Campanas and Mt Wilson.

The Deep Space CCD Atlas (two volumes) by John C. Vickers, P.O. Box 1292, Duxbury, MA 02331. Hundreds of laser printed amateur galaxy and Deep Sky images by Vickers; ideal for supernova patrolling.

The Colour Atlas of Galaxies by James D. Wray (1988). Cambridge University Press. Six hundred colour images from professional observatories.

The Hubble Atlas of Galaxies by Allan Sandage, first edition, 1961. Carnegie Institute of Washington DC. Reproductions of the original Mt Wilson and Palomar images.

Atlas of Galaxies by Allan Sandage and John Bedke (1988). NASA SP-496. 300 galaxy photographs from Las Campanas and Palomar Observatories.

The *Supernova Search Charts and Handbook* by R. Bryan and L. Thomson (1989). Cambridge University

Press. Over 300 galaxies in both hemispheres are represented. A bit dated nowadays, but still useful for the visual observer if you can track a copy down.

A CCD Atlas of Deep Sky Objects (CD-ROM only) by C. Buil and E. Thouvenot (1991). Sky Publishing. This was the first digital Deep Sky atlas but is still available. Other amateurs, such as John Vickers and Pedro Re, have produced their own Deep Space atlases in recent years.

The Photographic Atlas of the Stars by Arnold, Doherty and Moore. Institute of Physics Publishing. ISBN 0 7503 0378 6. Forty five plates, each spanning 55° × 38°, are the backbone of this book. The field of view is similar to that of the human eye, but with binocular-level magnitude limits.

Dealers, Bibliography and URLs

Equipment Suppliers

All of the following distributors and suppliers are highly recommended by the author:

For frank and detailed reviews of astronomical accessories, take out a subscription to *Sky & Telescope*.

- Sky & Telescope, PO Box 9111, Belmont MA 02178-9111, USA; telephone (US) (800) 253-0245.

For UK purchases of US-made equipment, the largest British distributor and agent for the leading US companies is:

- Broadhurst Clarkson and Fuller, Telescope House, 63 Farringdon Road, London EC1M 3JB.

The UK distributor for Celestron telescopes and accessories is:

- David Hinds Ltd; telephone (UK) (01442) 827768.

For world-wide mail order of Deep Sky filters and hypering kits, contact:

- Lumicon, 2111 Research Drive, Suites 4-5, Livermore, CA 94550, USA; telephone (US) (925) 447-9570.
- Pocono Mountain Optics, 104N P 502 Plaza, Moscow, PA 18444, USA; telephone (US) (717) 842-1500.

For Cyanogen Productions remarkable Comet Hyakutake video plus image-processing software contact:

- Cyanogen Productions, Inc., 25 Conover Street, Nepean, Ontario, K2G 4C3, Canada; telephone (US) (613) 225-2732.

For digital setting circles, motorised and quality focusers and other retro-fit accessories, contact:

- JMI, 810 Quail Street, Unit E, Lakewood, CO 80215, USA; telephone (US) (303) 233-5353.

For CCD autoguiders and cameras, contact:
- SBIG (Santa Barbara Instruments Group), 1482 East Valley Road, Suite #J601, Santa Barbara, CA 93108, USA; telephone (US) (805) 969-1851.

In the UK, for the British-made "Starlite Xpress" CCD cameras contact:
- FDE Ltd; telephone (UK) (01734) 342600.

In the UK, for Takahashi telescopes, Hi-Sis CCDs, QMips Software, Miyauchi Binoculars and Flip Mirrors, contact:
- True Technology Ltd, Woodpecker Cottage, Red Lane, Aldermaston, Berks, RG7 4PA.

Quality Binocular Mounting Suppliers

The GrandView Binocular Mount, GrandView Instruments, P.O. Box 278, Concord CA 94522, USA.

Sky Hook, Star Bound, 68 Klaum Ave., N. Tonawanda, NY 14120, USA.

Vista Binocular Guider, Vista Instrument Co., P.O. Box 1919, Santa Maria, CA 93456, USA.

Societies Worth Joining

The *British Astronomical Association*, Burlington House, Piccadilly, London W1V 9AG, UK. 3000 amateur astronomers, mainly based in the UK. Formed in 1890 and with an observational reputation second to none. Bi-monthly quality journal featuring papers on members, observations, news and historical research.

The Astronomer magazine. Secretary: Pete Meadows, 6 Chelmerton Avenue, Gt Baddow, Chelmsford, Essex, CM2 9RE, UK. (A team of 250 very active observers, especially so for variable star and comet observers, led by Guy Hurst.)

The Association of Lunar and Planetary Observers (ALPO). Membership Secretary, P.O. Box 16882, Memphis, TN 38186-0882, USA.

International Comet Quarterly, Mail Stop 18, Smithsonian, Astrophysical Observatory, 60 Garden Street, Cambridge, MA 02138. USA, (A magazine for the serious comet observer.)

Internet URLs

Undoubtedly the quickest way of getting up-to-date information on new astronomical products and astronomy in general is via the PC modem and the Internet. Most equipment manufacturers now have their own web page and many allow ordering via the Internet. The URLs below may well be the quickest way of finding and ordering the product you want.

URL	Content
http://www.Meade.com	Meade
http://www.austin.cc.tx.us/ astro-ES/AstroDesigns/index.htm	Meade User Group
http://www.austin.cc.tx.us/ astro-ES/AstroDesigns/MAPUG/ ArhvList.htm	Meade User Group
http://www.mailbag.com/users/ ragreiner	Useful Meade info.
http://www.celestron.com	Celestron homepage
http://ourworld.compuserve.com/ homepages/AVAastro/	Starlight Xpress
http://www.pcug.co.uk/~starlite/	Starlight Xpress
http://www.sbig.com/	SBIG CCDs
http://www.optecinc.com	Telecompressors
http://www.skypub.com/	Sky Publishing
http://www.skypub.com/testrept /testrept.shtml	Equipment reviews
http://www.users.dircon.co.uk/ ~truetech/	True Technology
http://www.bisque.com	Software Bisque
http://www.oriontel.com	Orion Scope Center
http://www.Astronomy-Mall.com	Pocono Mt. Optics
http://www.kendrick-studio.com	Kendrick Dew Zapper
http://www.lasermax-inc.com	Laser collimators
http://ourworld.compuserve.com/ homepages/johneroger	CCD astrometry
http://www.StarryMessenger.com	Used telescopes
http://www.ast.cam.ac.uk/~baa/	British Astronomical Association
http://www.demon.co.uk/ astronomer/	The Astronomer

http://www.astronomynow.com/	Astronomy Now
http://cfa-www.harvard.edu/cfa/ps/cbat.html	CBAT/IAU/ICQ
http://cfa-www.harvard.edu/cfa/ps/Headlines.html	CBAT Headlines
http://huey.jpl.nasa.gov/~spravdo/neat.html	Near Earth Asteroids
http://www.gsfc.nasa.gov/NASA_homepage.html	NASA
http://stdatu.stsci.edu/cgi-bin/dss_form	Palomar Sky Survey
http://www.seds.org/ftp/images/deepspace/gco	Galaxy Images
http://www.aspsky.org/html/resources/ngc.html	Galaxy Images
http://cfa-www.harvard.edu/cfa/ps/lists/RecentSupernovae.html	Recent supernovae
http://cfa-www.harvard.edu/cfa/ps/lists/Supernovae.html	All supernovae
http://www.queen.it/web4you/noprofit/isn/isn.htm	International Supernova Network
http://ourworld.compuserve.com/homepages/MartinMobberley/	Martin Mobberley!

Books

Handbook for Telescope Making by N.E. Howard. Faber and Faber. Reprinted 1970. ISBN 0 571 04680 0. A classic book on telescope making. Out of print but worth tracking down!

Amateur Astronomers Handbook by J.B. Sidgwick. Faber and Faber. Third edition, 1971. ISBN 0 571 04748 3. The 1970's telescope equipment "bible". Worth tracking down in second-hand bookshops or libraries.

Small Astronomical Observatories, edited by Patrick Moore. Springer-Verlag, 1996. ISBN 3-540-19913-6. Twenty-five observatories and how they were built – invaluable observatory experience captured in one book.

How to make a Telescope by J. Texereau. Willman-Bell, 1984. Another classic.

Building and Using an Astronomical Observatory by Paul Doherty. Patrick Stephens Ltd, 1986. ISBN 0-85059-808-7.

Out of print but worth tracking down. Excellent practical advice from a master observer.

Astrophotography with the Schmidt Telescope by Marx and Pfau. Cambridge University Press, 1992. ISBN 0521 395496. The history of the Schmidt telescope – a must for Schmidt camera fans.

The Messier Album by Mallas and Kreimer. Cambridge University Press, 1978. Two amateurs, experiences with observing and photographing the Messier objects using sketchpad and cold camera. Remarkable photographs from the 1960s.

Eyes on the Universe – The Story of the Telescope by Patrick Moore. Springer-Verlag, 1997. ISBN 3540 761640. A slimline book summarising the story of the telescope. Issued to celebrate Patrick Moore's 40th Anniversary "Sky at Night" programme.

Exploring the Night Sky with Binoculars by Patrick Moore. Cambridge University Press. 3rd edition, 1996. ISBN 0-521-5538-8.

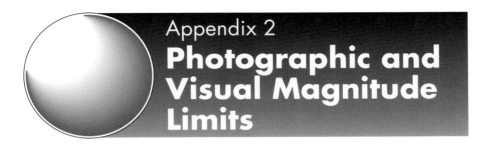

Photographic and Visual Magnitude Limits

I have been involved in astrophotography for twenty years and during that time have amassed a great deal of data on the limiting photographic and CCD magnitudes achievable with various instruments under various conditions. This information, along with visual magnitude limits, are reproduced below.

The standard formula for predicting the limiting visual magnitude of a telescope is:

$$\text{Lim } m_v = 2 + 5\log_{10}D$$

where D is the diameter of the clear aperture of the telescope expressed in millimetres. The values in the table below are derived from this formula:

Aperture (mm)	Limiting m_v
60	10.8
76	11.4
102	12.0
152	12.9
203	13.5
254	14.0
305	14.4
356	14.8
406	15.0
457	15.3
508	15.5

It should be stressed that to achieve these results, dark skies, good optics and a dark-adapted eye are essential prerequisites. In addition, high magnifications will increase the contrast and improve the observer's limiting magnitude. Keen-eyed observers can often see to a magnitude fainter than the limits quoted here. Conversely, some observers may find that stars a magnitude brighter than those quoted are nearer to their personal limit. When first going outside, with dark adaption of only a few minutes, one can typically only see stars that are 3–4 magnitudes brighter than those quoted here.

I have achieved the following limiting magnitudes with T-Max 400, Hypered Kodak 2415 and Interline and Frame Transfer CCDs using a 360 mm f/5 Newtonian from a dark, rural site:

Exposure (mins)	Kodak T-Max 400	Hypered 2415	Interline CCD	Frame Trans. CCD
1	14.8	14.4	17.0	17.8
2	15.4	15.2	17.7	18.5
4	16.0	15.9	18.4	19.2
8	16.5	16.6	19.0	19.8
15	16.9	17.3	19.5	20.2
30	17.2	18.0	20.0	20.6

As already mentioned in the "Astrophotography" section in Chapter 7, an 85 mm lens at f/2 and 1-minute exposures on T-Max 400, from a rural site in the UK reach a limiting magnitude of around 11.0. From altitude in the clear, dark skies of Tenerife, I have achieved magnitude 12.8 in 3 minutes with an 85 mm aspheric lens at f/1.2, using hypered 2415. You should, incidentally, realise that even from a dark country site in the UK, the skies are far brighter than at altitude on Tenerife. Hypered 2415 will be badly fogged in under two minutes at f/2, even from a rural UK site, especially if objects at low altitude are being imaged.

At f/5, however, exposures of up to 30 minutes can be attempted from dark country sites without excessive fogging. The instrument f/ratio is critical where film fogging is concerned.

Index